AF292831

Progress in Nonlinear Differential Equations and Their Applications
Volume 9

Editor

Haim Brezis
Université Pierre et Marie Curie
Paris
and
Rutgers University
New Brunswick, N.J.

Editorial Board

Manuel D.P. Monteiro Marques

Differential Inclusions in Nonsmooth Mechanical Problems

Shocks and Dry Friction

1993

Springer Basel AG

Manuel D. P. Monteiro Marques
Centro de Matemática e Aplicações Fundamentais
Av. Prof. Gama Pinto, 2
1699 Lisboa Codex
Portugal

A CIP catalogue record for this book is available from the Library of Congress,
Washington D.C., USA

Deutsche Bibliothek Cataloging-in-Publication Data
Monteiro Marques, Manuel D. P.:
Differential inclusions in nonsmooth mechanical problems :
shocks and dry friction / Manuel D. P. Monteiro Marques. –
Basel ; Boston ; Berlin : Birkhäuser, 1993
  (Progress in nonlinear differential equations and their applications ;
  Vol. 9)
NE: GT

ISBN 978-3-0348-7616-2   ISBN 978-3-0348-7614-8 (eBook)
DOI 10.1007/978-3-0348-7614-8

Contents

Chapter 5

Further Applications and Related Topics

Introduction

This book is devoted to the sudy of some differential inclusions motivated by Mechanics and of existence results for the dynamics of systems with inelastic shocks, with or without friction. This ensures a certain unity of subject, techniques and applications, at the price of not including some earlier works [Mon 1-4] .

In the introductory Chapter 0, several essential mathematical tools (either recent or recently rediscovered) are presented. Mainly they concern functions of bounded variation defined in real intervals (derivation of Stieltjes measures, compactness results, convergence in the sense of graphs) and geometrical inequalities.

In Chapters 1 and 2, Moreau's sweeping process is considered; this is a first-order differential inclusion

$$(1) \qquad\qquad -du\,\varepsilon\; N_{C(t)}(u(t))\;,$$

where the right-hand side is the outward normal cone to a convex set $C(t)$ at a point $u(t)$ of a real Hilbert space H. This abstract formulation encompasses several practical situations in Mechanics, such as water falling in a cavity [Mor 1], the dynamics of systems with perfect unilateral constraints [Mor 15,16], plasticity and the evolution of elastoplastic systems [Mor 3, 17].

In Chapter 1, based on [Mon 5], it is assumed that the moving convex set, i. e., the multifunction $t \to C(t)$ has right-continuous bounded variation (in short, it is said to be rcbv). Then a unique rcbv solution of (1) is known to exist, for any prescribed admissible initial value. It is proved here that even in this case the regularization technique of Yosida (or Moreau-Yosida) still gives a sequence of absolutely continuous functions that converge to the solution. Since the latter may be discontinuous, in general the convergence is not uniform but "in the sense of filled-in graphs". This numerically meaningful notion of convergence, previously introduced in Chapter 0, looks bound to gain importance, as more and more discontinuous problems are currently being treated. Several other technical results on convergence are given.

Chapter 2, which is based on [Mon 6, 7], deals with the case of a convex set $C(t)$ which is not necessarily of bounded variation in time but has nonempty interior instead. The existence of a cbv (continuous with bounded variation), respectively of an rcbv solution, is proved for two cases: a) the multifunction $t \to C(t)$ is continuous in the sense of Hausdorff distance and the Hilbert space has arbitrary dimension; b) the convex set is lower semicontinuous from the right and H is finite dimensional. Although the starting point is the well-known discretization technique of Moreau (the catching-up algorithm) the proofs are substantially different and recent technical advances are needed. Chapter 2 also contains a random or parametrized version of inclusion (1), as well as some *a priori* bounds and a study of the dependence on the data.

Chapter 3, partially based on [Mon 8], is concerned with the problem of the dynamics of a mechanical system with a finite number of degrees of freedom, subject to a unique unilateral constraint and experiencing inelastic shocks. The constraint is expressed by an inequality $f(q) \leq 0$, defining a region L. In the frictionless case, the synthetic modern formulation of Moreau [Mor 11] gives rise to the second-order differential inclusion:

$$(2) \qquad p(t, q(t))\, dt - du \in N_{V(q(t))}(u(t)) \, ,$$

where $u = \dot{q}$ is the rcbv velocity, p is a force field and $V(q(t))$ is the (convex) tangent cone to L at the position $q(t)$. The existence of a solution to (2) is proved by constructing a suitable sequence (q_n) of approximants of the motion and extracting a uniformly convergent subsequence. The rather long proof uses some of the techniques of Chapter 2, thus explaining a somewhat weak regularity assumption on L (its boundary only needs to be $\mathcal{C}^1$).

With a similar approach, it is shown that there exists at least one solution to the problem with Coulomb's isotropic friction, expressed by a differential inclusion that can be written as:

$$(3) \qquad -u(t) \in \mathrm{proj}_{T(q(t))} \, N_{C(q(t))}(du - p(t, q(t))\, dt) \, ,$$

where $C(q(t))$ is the friction cone and $T(q(t))$ is the tangent hyperplane to L at $q(t)$. In this currently active research field there remain many open problems, e. g., to find conditions ensuring uniqueness of the solution and to prove existence in the presence of more than one unilateral constraint.

In Chapter 4, it is studied a problem derived from (2) and involving what may be called externally induced collisions (or shocks or contacts). An "activation multifunction" $t \to A(t)$ is given, with values in a finite set $\{1, ..., \nu\}$ and which selects among the constraints $f_1(q) \le 0$, ... , $f_\nu(q) \le 0$ those that are active at time t. The corresponding differential inclusion is the following:

$$(4) \qquad p(t, q(t))\, dt - du\, \varepsilon\, N_{W(A,\, t,\, q(t))}(u(t)) \, ,$$

where $W(A, t, q(t))$ is a convex set, lower semicontinuous with respect to t, associated to A. By exploiting the introduced techniques and fixed point theorems, the existence of solution is obtained, although not the uniqueness. The relation between the problems of inelastic shocks and of induced collisions is also discussed.

In the final Chapter 5, some directions of research that can be linked to the present exposition are reviewed. A short presentation is made of three works: 1) existence for some second-order differential inclusions by C. Castaing; 2) the approximation techniques for differential inclusions recently developed by M. Valadier and A. Gavioli; and 3) application of lower semicontinuous differential inclusions to quasi-statics by Chraibi. In section 5.4 the list of references on related subjects is expanded.

It is assumed that the reader is familiar with basic concepts of Real Analysis, General Topology and some Functional Analysis. In particular, a preliminary knowledge of evolution equations and of set-valued analysis is advisable. Nevertheless, some effort was done in order to render the reading as self-contained as possible (in what goes beyond standard textbooks).

I am deeply grateful to Prof. Jean Jacques Moreau (Montpellier), who introduced me to the field and suggested most of the research problems, for his warm-hearted support and enlightened scientific guidance.

I am also grateful to Profs. Charles Castaing and Michel Valadier (Montpellier) for stimulating discussions on the subject; to Prof. Panagiotis Panagiotopoulos (Thessaloniki and Aachen) for encouraging me to write the book; to E. F. Beschler (Birkhäuser) and to my colleagues Owen Brison, João Martins (IST), Augusto Franco de Oliveira, Luís Saraiva and Luís Trabucho for helping improve my English; to Bento Louro and Armando Machado for their help with hardware and software; and to colleagues and friends at the Mathematical Department of the Science Faculty FCUL (Faculdade de Ciências da Universidade de Lisboa) and at the Research Center CMAF (Centro de Matemática e Aplicações Fundamentais) for creating such a warm and joyful work environment, even under sometimes adverse circumstances.

I am indebted to several institutions that supported this research at different stages throughout the years, namely to FCUL, to the French Government, to Fundação Calouste Gulbenkian (Lisboa) and to the former INIC (Instituto Nacional de Investigação Científica).

I have great pleasure in thanking the publishers for fruitful collaboration, in particular by Dr. Thomas Hintermann.

Lisboa, March 1993Manuel D. P. Monteiro Marques

Chapter 0

Preliminaries

0.1. Functions of bounded variation

In this section, we gather some information on functions of bounded variation of a single real variable. For a detailed study, we refer to [Mor 10] which we use extensively.

Let I be a nonempty real interval and H be a real Hilbert space, with norm $\| \; \|$ (although for part of the results we could take, more generally, a real Banach space or a metric space). We consider functions $u: I \to H$. For every t different from the possible right end of the interval, $u^+(t)$ denotes the right-limit

$$u^+(t) = \lim_{s \to t, \, s > t} u(s) \, ,$$

if it exists. The left-limit $u^-(t)$ is defined similarly. If t is the right endpoint of I, we agree to write $u^+(t) = u(t)$. If t is the left endpoint, we write $u^-(t) = u(t)$; if $u(t)$ is an initial datum, say an initial velocity, this convention means that $u(t)$ conveys some abridged information on the system history before t.

If they are defined everywhere in an open interval, then u^+ is a right-continuous function and u^- is a left-continuous function, i.e., $(u^+)^+ = u^+$ and $(u^-)^- = u^-$. Also, $(u^-)^+ = u^+$ and $(u^+)^- = u^-$.

The *variation* of u in a subinterval J of I is the nonnegative extended real number

$$\text{(1)} \qquad \mathrm{var}(u, J) = \sup \, \sum_{i=1}^{n} \| u(t_i) - u(t_{i-1}) \| \, ,$$

where the supremum is taken in $[0, +\infty]$ with respect to all the finite sequences $t_0 < t_1 < \ldots < t_n$ of points of J (n is arbitrary). For any singleton, $\mathrm{var}(u, \{t\}) = \sup \emptyset = 0$. More generally, $\mathrm{var}(u, J) = 0$ if and only if u is constant on J. We also write $\mathrm{var}(u; a, b)$ instead of $\mathrm{var}(u; [a, b])$, with the property that $\mathrm{var}(u; a, c) = \mathrm{var}(u; a, b) + \mathrm{var}(u; b, c)$.

The function $u: I \to H$ is said of *bounded variation on* I if and only if $\mathrm{var}(u, I) < +\infty$. We often say that u is bv or is a bv function and write $u \in bv(I, H)$. The set of functions of bounded variation is a linear space, since

$\mathrm{var}(\lambda u + \beta v; I) \leq |\lambda|\, \mathrm{var}(u, I) + |\beta|\, \mathrm{var}(v, I)$. If u is bv, the above addition rule implies that there exists a nondecreasing function $V_u : I \to \mathbb{R}$ (called a *variation function* of u) such that

$$(2) \qquad\qquad \forall [a, b] \subset I : \quad \mathrm{var}(u; a, b) = V_u(b) - V_u(a) \ ;$$

for instance, $V_u(a) = \mathrm{var}(u; c, a)$ (resp. $-\mathrm{var}(u; a, c)$) if $a \geq c$ (resp. if $a < c$) where $c \varepsilon I$ is fixed. Conversely, (2) implies that u is *of locally bounded variation* (*lbv* for short), that is, u has finite variation on any compact subinterval of I. If u is Lipschitz-continuous (resp. absolutely continuous) then V_u is Lipschitz-continuous (resp. absolutely continuous) and conversely. In fact, assuming for example that V_u is Lipschitz-continuous with constant k, then

$$\| u(b) - u(a) \| \ \leq\ \mathrm{var}(u; a, b) = V_u(b) - V_u(a) \ \leq\ k\,| b - a | \ .$$

A bv function u defined on $[0, T]$ possesses left-limits in $]0, T]$ and right-limits in $[0, T[$. A variation function V_u is left, or right or simply continuous at t if and only if the same holds with the function u. This is a consequence of formulas such as (see [Mor 10] Prop.4.3):

$$\lim_{s \uparrow t} \| u(s) - u(t) \| = \lim_{s \uparrow t} \mathrm{var}(u; s, t) = V_u(t) - V_u^-(t) \, .$$

As it is well known that a nondecreasing real function, such as V_u, has a countable set of discontinuity points, the same is true for any bv function u.

It can be shown that, if u is bv, then u^- and u^+ are both bv functions, with for instance

$$\mathrm{var}(u^- ; a, b) \leq \ V_u^-(b) - V_u^-(a) \ .$$

We now define the integral with respect to the differential measure of u of a continuous real function ϕ on I whose support, relative to I, is compact; we may write $\phi \varepsilon \mathcal{K}(I)$. We denote by $\mathcal{P}$ the set of finite subsets of I. If P belongs to $\mathcal{P}$, its elements, which we call *nodes*, may be ordered, say P: $t_0 < t_1 < ... < t_n$. The elements of $\mathcal{P}$ are themselves ordered by inclusion: $P \geq Q$ iff $P \supset Q$. Hence, $\mathcal{P}$ becomes a directed set and we can take limits of families indexed on $\mathcal{P}$, that is of *nets* or generalized sequences. Let us say that θ is an *intercalator* if it associates with every $P \varepsilon \mathcal{P}$ and every i an element $\theta_{P,i} \varepsilon [t_{i-1}, t_i]$. If u is a bv function and θ is an intercalator, we may define the (Riemann-Stieltjes) sums, indexed by $P \varepsilon \mathcal{P}$,

$$(3) \qquad\qquad S(\phi, P, \theta ; u) = \sum_{i=1}^{n} \phi(\theta_{P,i})(u(t_i) - u(t_{i-1})) \, ,$$

and prove that they converge to a limit, which is independent of the intercalator; moreover, the convergence is uniform with respect to θ. We define:

$$(4) \qquad \int \phi\, du = \lim_{P \in \mathcal{P}} S(\phi, P, \theta\,;\, u)\,.$$

If $[a, b]$ is a compact subinterval of I containing the support of ϕ, then it follows from the above definition that:

$$(5) \qquad \left\| \int \phi\, du \right\| \leq \max |\phi|\ \operatorname{var}(u\,;\, a, b)\,.$$

This shows that the linear map $\phi \to \int \phi\, du$ from $\mathcal{K}(I)$ to H is continuous and so is a vector measure on I, in the sense of Bourbaki. This H-valued measure is called the **differential measure** or **Stieltjes measure** of the function $u \in bv(I, H)$ and denoted by du. To approximate the value of the integral (4) we may use the following: let $\epsilon > 0$ and let an interval $[a, b]$ contain the support of ϕ; by uniform continuity of ϕ on $[a, b]$, we choose P with nodes so close that $|\phi(t) - \phi(s)| \leq \epsilon$, for every i and every s, t in $[t_{i-1}, t_i]$; then, $P' \supset P$ with $P' = (t'_i)$ implies that ([Mor 10], (6.10)):

$$(6) \qquad \left\| \int \phi\, du - \sum_{i=1}^{n} \phi(\theta_{P',i})\,(u(t'_i) - u(t'_{i-1})) \right\| \leq \epsilon\ \operatorname{var}(u\,;\, a, b)\,.$$

The map $u \to du$ is linear, with values in the real linear space of all H-valued measures. If the vector function u is a constant, then $du = 0$. The converse is true if u is right-continuous (alternatively, left-continuous) in the interior of I.

If u is a step function, there is $P \in \mathcal{P}$ such that u is constant on every interval of P. In this case, the integral (4) is given by

$$(7) \qquad \int \phi\, du = \sum_{k=1}^{m} \phi(t_k)\,(u^+(t_k) - u^-(t_k))\,.$$

Only the nodes contained in the support of ϕ need to be considered in (7). In other words, the differential measure du equals the sum of a finite collection of point (Dirac) measures, placed at the discontinuity points of u, their values being the respective jumps of u:

$$(8) \qquad du = \sum_{k} (u^+(t_k) - u^-(t_k))\,\delta_{t_k}\,.$$

This is also true for *local* step functions: in that case we get in (8) a locally finite collection of point measures. The formula also shows that we may change the value of u at an interior point, say t_k, without changing the differential measure.

We next recall some concepts and properties of general vector measures.

A vector measure m on I is *majorable* iff there exists a nonnegative real measure μ on I such that, for every $\phi \varepsilon \mathcal{K}(I)$:

$$(9) \qquad \left\| \int \phi \, dm \right\| \leq \int |\phi| \, d\mu \, .$$

By the usual decomposition of a function as a difference of nonnegative functions, it suffices to verify (9) for nonnegative functions $\phi \varepsilon \mathcal{K}_+(I)$. If H has finite dimension, then any H-valued measure is majorable.

Consider a majorable H-valued measure, where H is a (real) Hilbert space. For fixed $x \varepsilon H$, the linear map $\phi \to x . \int \phi \, dm$ from $\mathcal{K}(I)$ to $\mathbb{R}$ is a real measure, denoted by $x . m$. If $\|x\| \leq 1$, we have:

$$(10) \qquad \forall \phi \varepsilon \mathcal{K}_+(I): \quad x . \int \phi \, dm \leq \|x\| \int \phi \, d\mu \leq \int \phi \, d\mu \, ,$$

so that $x . m \leq \mu$ in the sense of the ordering of real measures. The supremum of the measures $x . m$ with $\|x\| \leq 1$ is then a nonnegative real measure $\leq \mu$; it is called the *modulus* or *absolute value* of m and it is denoted by $|m|$ or $|dm|$. In case $H = \mathbb{R}$, this is the usual absolute value of a real signed measure. By definition, $x . \int \phi \, dm \leq \int \phi \, |dm|$; taking $x = \| \int \phi \, dm \|^{-1} \int \phi \, dm$, we have

$$(11) \qquad \left\| \int \phi \, dm \right\| \leq \int \phi \, |dm| \, , \quad \forall \phi \varepsilon \mathcal{K}_+(I) \, .$$

Consequently, (9) and (10) hold with $d\mu$ replaced by $|dm|$.

A real function $h : I \to \mathbb{R}$ is *scalarly integrable* with respect to m if it is integrable with respect to all real measures $x . m$, with $x \varepsilon H$. The *integral* of h is then defined as the element $\int h \, dm$ of H for which

$$(12) \qquad \int h \, d(x . m) = x . \int h \, dm \, , \quad \forall x \varepsilon H,$$

if it exists. This is always the case if m is majorable and $h \varepsilon \mathcal{L}^1(I, |m|; \mathbb{R})$ and then:

$$(13) \qquad \left\| \int h \, dm \right\| \leq \int |h| \, |dm| \, .$$

An extension of the *dominated convergence theorem* asserts the following: let $(h_n) \subset \mathcal{L}^1(I, |m|; \mathbb{R})$ converge $|m|$-almost everywhere to h and let $|h_n(t)| \leq g(t)$ for $|m|$-almost every t in I, with $g \varepsilon \mathcal{L}^1(I, |m|; \mathbb{R}^+)$; then the limit h is also an integrable function with respect to $|m|$ and

$$(14) \qquad \lim_n \int h_n \, dm = \int h \, dm \, ,$$

in the sense of the norm of H.

An important example of a majorable H-valued measure is the product of a real measure μ and a locally integrable vector function $m'_\mu \varepsilon \mathcal{L}^1_{loc}(I,\mu;H)$. We denote it by $m = m'_\mu \mu$ or $dm = m'_\mu \, d\mu$ and we define:

$$(15) \qquad \int \phi \, dm = \int \phi \, m'_\mu \, d\mu \, , \quad \forall \phi \, \varepsilon \, \mathcal{K}(I) \, .$$

In these circumstances, we say that m admits m'_μ as a *density* relative to μ. Of course, we may replace m'_μ by any μ-equivalent H-valued function: a density of m may be considered as a uniquely defined element of $L^1_{loc}(I,\mu;H)$. If μ is nonnegative, the modulus measure is given by $\quad |\, m'_\mu \mu \,| = \|\, m'_\mu \,\| \, \mu \, .$

Since any Hilbert space has the so-called Radon-Nikodym property, it is true that every majorable measure m is of the above form (15) with μ equal to the modulus measure of m; in other words, m has a density with respect to $|\, m \,|$. For instance, let $m = a\delta_t$, with $a \varepsilon H \setminus \{0\}$, so that $\int h \, dm = h(t)\, a$; then $\mu = |\, m \,| = \|\, a \,\| \, \delta_t$ and $m'_\mu = \|\, a \,\|^{-1} a$.

In the special case of a Stieltjes measure du of a bv function u it is known that du is majorable and that, in the sense of the ordering of real measures:

$$(16) \qquad\qquad |\, du \,| \leq dV_u \, .$$

In fact, for $\phi \, \varepsilon \, \mathcal{K}_+(I)$, we have $\|\, S(\phi, P, \theta; u) \,\| \leq \sum \phi(\theta_{P,i}) \|\, u(t_i) - u(t_{i-1}) \,\|$
$\leq \sum \phi(\theta_{P,i}) (\, V_u(t_i) - V_u(t_{i-1})) = S(\phi, P, \theta; V_u)$. This implies that, in the limit with respect to P, $\|\, \int \phi \, du \,\| \leq \int \phi \, dV_u$, whence $\int \phi \,|\, du \,| \leq \int \phi \, dV_u$, by definition of $|\, du \,|$. Notice that, in general, the equality does not hold in (16): take u constant except at an interior point of I and obtain $|\, du \,| = 0 \neq dV_u$. However, if u has *aligned jumps*, i.e., if for every $t \varepsilon I$ the value $u(t)$ belongs to the line segment with endpoints $u^-(t)$ and $u^+(t)$, then

$$(17) \qquad\qquad |\, du \,| = dV_u \, .$$

Conversely, (17) implies that u has aligned jumps, because H has a strictly convex norm. It is clear that (17) applies to right-continuous bv functions. The measure $|\, du \,|$ is also called the *measure of total variation* of u (compare (17) and (2) to obtain $\int |\, du \,| = \mathrm{var}(u)$).

Like real measures, vector measures are uniquely determined by the values they take on compact subintervals, that is by the integrals of the respective characteristic functions. These are usually denoted by the letter χ: $\chi_A(x) = 1$, if $x \varepsilon A$ and 0 otherwise. For Stieltjes measures, we have:

$$(18) \qquad \int_{[a,\,b]} du = \int \chi_{[a,b]}\, du = u^+(b) - u^-(a) \,,$$

with the particular case of a singleton:

$$(19) \qquad \int_{\{a\}} du = u^+(a) - u^-(a) \,.$$

This follows by approaching the characteristic function by a sequence of continuous functions and applying the dominated convergence theorem. Since $\chi_{[a,\,b]} = \chi_{\{a\}} + \chi_{]a,\,b]}$, (18) and (19) imply that

$$(20) \qquad \int_{]a,\,b]} du = u^+(b) - u^+(a) \,.$$

By similar reasonings,

$$(21) \qquad \int_{[a,\,b[} du = u^-(b) - u^-(a) \,,$$

$$(22) \qquad \int_{]a,\,b[} du = u^-(b) - u^+(a) \,.$$

A consequence of this and of $(u^+)^+ = u^+$ and $(u^-)^- = u^-$ is that $du^+ = du$ (if u is right-continuous at the left end of I) and $du^- = du$ (if u is left-continuous at the right end of I).

Let J be a nonempty subinterval of I. We denote by u_J the restriction of u to J; $d(u_J)$ is the differential measure of u_J. Let $\overline{\phi}$ be the extension of $\phi \in \mathcal{K}(J)$ to I with zero value outside J (notice that $\overline{\phi}$ belongs to $\pounds^1(I, |\,du\,|; \mathbb{R})$). The *measure induced by du* on J is denoted by $(du)_J$ and has the definition:

$$(23) \qquad \int \phi\, (du)_J = \int \overline{\phi}\, du \,, \quad \forall \phi \in \mathcal{K}(J) \,.$$

It can be shown that $(du)_J$ equals the sum of $d(u_J)$ and the following measures on J: (a) the measure $(u(a) - u^-(a))\delta_a$, if J contains its left end a; (b) the measure $(u^+(b) - u(b))\delta_b$ if J contains its right end b. The conventions about endpoints of I apply here. In the particular case of a singleton $J = \{a\}$ ($a = b$) we have $(du)_J = (u^+(a) - u^-(a))\delta_a$, whereas $d(u_J) = 0$. While $(du)_J \neq d(u_J)$ may look inconvenient, everything runs smoothly with the variation: $\mathrm{var}(u, J) = \mathrm{var}(u_J, J)$.

The Bourbaki construction of a Z-valued measure through a bilinear map $\Phi \colon X \times Y \to Z$ involving three Banach spaces can be specialized to the case where $\Phi \colon H \times H \to \mathbb{R}$ is the scalar product $\Phi(x, y) = x \cdot y$. If m is a majorable H–valued measure, we start by defining the integral $\int f . dm$ as a linear continuous function of $f \in \pounds^1(I, |\,m\,|; H)$ with values in $\mathbb{R}$. Since the linear space generated by functions of the form

$$(24) \qquad f = r\, a \quad , \quad a \varepsilon H \,, \quad r \varepsilon \mathcal{L}^1(I, \,|\, m\,| \,; \mathbb{R}) \,,$$

is dense in $\mathcal{L}^1(I, \,|\, m\,| \,; H)$, the formula

$$(25) \qquad \int (r\, a) \,.\, dm := a \,.\, \int r\, dm$$

defines uniquely the integral on the left-hand side. Also, the inequality

$$(26) \qquad \left|\, \int f \,.\, dm \,\right| \; \leq \; \int \,\|f\| \;\,|\, dm\,| \,,$$

is easily proved for functions of the form (24) and extends to integrable f.

For $g \varepsilon \mathcal{L}^1_{loc}(I, \,|\, m\,| \,; H)$, we define a real measure $g\,.\,m$ or $g\,.\,dm$ by the following:

$$(27) \qquad \forall \phi \varepsilon \mathcal{K}(I): \quad \int \phi\,(g\,.\,dm) := \int (\phi g) \,.\, dm \,;$$

notice that ϕg belongs to $\mathcal{L}^1(I, \,|\, m\,| \,; H)$ so that the integral on the right-hand side of (27) exists. This new measure is majorable and we have

$$(28) \qquad |\, g\,.\,m\,| \; \leq \; \|\, g\,\| \;\,|\, m\,| \,,$$

in the sense of measures.

If the vector measure m has a density $m'_\mu \varepsilon \mathcal{L}^1_{loc}(I, \mu\,; H)$ with respect to a real measure $\mu \geq 0$ ($m = m'_\mu\, \mu$), then a calculation rule is available:

$$(29) \qquad g\,.\,m = (g\,.\,m'_\mu)\,\mu \quad , \quad \forall g \varepsilon \mathcal{L}^1_{loc}(I, \,|\, m\,| \,; H) \,.$$

An equivalent assumption on g is that $g\,\|\, m'_\mu\,\| \varepsilon \mathcal{L}^1_{loc}(I, \mu\,; H)$, because $|\, m\,| = \|\, m'_\mu\,\|\, \mu$. By (27) and (29), for every continuous function with compact support ϕ:

$$(30) \qquad \int \phi\,(g\,.\,dm) := \int (\phi\, g)\,.\,dm = \int \phi\,(g\,.\,m'_\mu)\, d\mu \,.$$

In fact, (30) still holds if $\phi\, g$ is replaced by any integrable vector function f:

$$(31) \qquad \int f \,.\, dm = \int (f\,.\,m'_\mu)\, d\mu \,, \quad \forall f \varepsilon \mathcal{L}^1(I, \,|\, m\,| \,; H) \,.$$

To prove (31), we only need to consider functions of the form (24). By (25), $\int (r\, a)\,.\,dm = a\,.\,\int r\, dm = a\,.\,\int r\, m'_\mu\, d\mu$, so that

$$\int (r\, a)\,.\,dm = \int [a\,.\,(r\, m'_\mu)]\, d\mu = \int (r\, a\,.\,m'_\mu)\, d\mu \,.$$

Notice that every H-valued measure has a density with respect to its modulus measure. Hence, we might alternatively take (31), with $\mu = |\, m\,|$, as a definition of the integral in its left-hand side. As an example of the integral, take $m = b\,\delta_t$ $(b \in H)$. Then, $g.(b\,\delta_t) = (g(t).b)\,\delta_t$ for every $g: I \to H$, meaning that $\int \phi\,(g.dm) = \phi(t)\,(g(t).b)$.

If u and v are functions of bounded variation from I to H then $u.v: I \to \mathbb{R}$ has bounded variation too (this results from the estimate $|u(t).v(t) - u(s).v(s)| \leq \|u\|_\infty \|v(t) - v(s)\| + \|v\|_\infty \|u(t) - u(s)\|$ which leads to $\mathrm{var}(u.v, I) \leq \|u\|_\infty \mathrm{var}(v, I) + \|v\|_\infty \mathrm{var}(u, I)$). Its differential measure is given by the formulas:

$$(32) \qquad d(u.v) = v^- . du + u^+ . dv = v^+ . du + u^- . dv = \frac{v^+ + v^-}{2} . du + \frac{u^+ + u^-}{2} . dv ;$$

notice that these are meaningful vector measures, because bv functions such as u^+, u^-, v^+ and v^- are uniform limits of sequences of step-functions, hence they are integrable with respect to any measure on I. Notice also that the first equality implies the second one, because $u.v = v.u$ and because the roles of u and v may be interchanged; on the other hand, the third formula follows by taking the half-sum of the first two. The general idea of the proof is to use an approximation argument to reduce the study to the case of (local) step-functions u and v. Then, a partition of I can be found such that u, v and $u.v$ are constant in each of its members, which are either intervals not containing their ends or singletons (called nodes). We show that the first two measures in (32) take the same value in each $[a, b]$ contained in I. If $t_1 < ... < t_n$ are the nodes of the partition belonging to that subinterval, then we have

$$(33) \qquad \int_{[a, b]} d(u.v) = \sum_{i=1}^{n} [u^+(t_i) . v^+(t_i) - u^-(t_i) . v^-(t_i)] ,$$

given that $u.v$ is constant on members of the partition. On the other hand,

$$\int_{[a, b]} v^- . du + \int_{[a, b]} u^+ . dv = \sum_{i=1}^{n} v^-(t_i) . [u^+(t_i) - u^-(t_i)]$$

$$+ \sum_{i=1}^{n} u^+(t_i) . [v^+(t_i) - v^-(t_i)] ,$$

equals the right-hand side of (33) and this proves the result for step-functions.

The formulas for differentiating a scalar product generate formulas of *integration by parts* such as

$$(34) \qquad \int_{[a, b]} u^+ . dv = u^+(b) . v^+(b) - u^-(a) . v^-(a) - \int_{[a, b]} v^- . du .$$

When $u = v$, (32) yields a very useful equality:

$$(35) \qquad d(u^2) = (u^+ + u^-) . du .$$

If u is bv and continuous, then $d(u^2) = 2u . du$ has the familiar aspect of a chain rule. A general result by Moreau gives for the case of a scalar product:

$$(36) \qquad 2\,u^- . \, du \;\leq\; d(u^2) \;\leq\; 2\,u^+ . \, du \;.$$

To prove this, in view of (35) and of the approximation technique it suffices to show that $u^+ . \, du \geq u^- . \, du$, for every step-function u; with nodes chosen as above, we have in fact:

$$\int_{[a,b]} (u^+ - u^-) . \, du \;=\; \sum_{i=1}^{n} [u^+(t_i) - u^-(t_i)] . [u^+(t_i) - u^-(t_i)] \;\geq\; 0 \;.$$

Let us end this section with the Moreau-Valadier extension to Banach-valued measures of Jeffery's result on the derivation of real measures [Jef].

Theorem 1.1 [Mor-Val 3] *Let I be a real interval, X a real Banach space, $d\mu$ a nonnegative (Radon) measure on I and $d\nu$ an X-valued measure on I admitting a density $\frac{d\nu}{d\mu} \in L^1_{loc}(I, d\mu; X)$. Then, for $d\mu$-almost every t in I:*

$$(37) \qquad \frac{d\nu}{d\mu}(t) \;=\; \lim_{\epsilon \downarrow 0} \frac{d\nu([t, t+\epsilon])}{d\mu([t, t+\epsilon])} \;=\; \lim_{\epsilon \downarrow 0} \frac{d\nu([t-\epsilon, t])}{d\mu([t-\epsilon, t])} \;.$$

Proof. [Mor-Val 3] Let us prove for example the first equality.

If $t = t_r$ (the possible right end of I) and $d\mu(\{t_r\}) = 0$, then we let t_r belong to the excluded $d\mu$-null subset. If $d\mu(\{t_r\}) > 0$ then, for any positive ϵ and with a natural convention, we have $d\mu([t_r, t_r + \epsilon]) = d\mu([t_r, t_r + \epsilon] \cap I) = d\mu(\{t_r\})$ and similarly $d\nu([t_r, t_r + \epsilon]) = d\nu(\{t_r\})$, while precisely

$$\frac{d\nu}{d\mu}(t_r) = \frac{d\nu(\{t_r\})}{d\mu(\{t_r\})} \;,$$

thus proving the formula in this case.

If $t \in I$ is not the right end, then for ϵ small enough we have $[t, t+\epsilon] \subset I$. To avoid dividing by zero, we must exclude those t belonging to

$$I_r = \{\, t \in I : \; d\mu([t, t+\alpha]) = 0 \text{ for some } \alpha > 0 \,\} \;,$$

which is a $d\mu$-null set. To see this, let I_0 be the greatest open subset of I in restriction to which $d\mu$ vanishes and write I_0 as a countable union of disjoint subintervals $J_n = \,]t_n, s_n[$. Then I_r is the union of I_0 and possibly some of the left ends t_n; since $t_n \in I_r \setminus I_0$ implies $d\mu(\{t_n\}) = 0$, by definition of I_r, it follows that $d\mu(I_r) = d\mu(I_0) = 0$.

Let h be a representative of the density $d\nu/d\mu$ and Z be a separable subspace of X, with dense sequence (z_n), such that $h(t) \in Z$ for $d\mu$-almost every t. We may apply *Jeffery's theorem* to every nonnegative measure $d\nu_n := \| h(.) - z_n \| \, d\mu$, obtaining a $d\mu$-null set $N_n \subset I$ such that:

$$(38) \qquad \| h(t) - z_n \| \;=\; \frac{d\nu_n}{d\mu}(t) \;=\; \lim_{\epsilon \downarrow 0} \frac{d\nu_n([t, t+\epsilon])}{d\mu([t, t+\epsilon])} \;, \quad \forall t \notin N_n \;.$$

The union of all the sets N_n with the set $\{t: h(t) \notin Z\}$ is a $d\mu$-null set N and we prove that

$$(39) \qquad \forall\, t \notin N : \frac{d\nu([t, t+\epsilon])}{d\mu([t, t+\epsilon])} \to h(t) \ , \ \text{as} \ \epsilon\!\downarrow\!0 .$$

Fix $t \notin N$ and an arbitrary $\eta > 0$. We choose n such that $\| z_n - h(t) \| \leq \eta$. Writing I_ϵ instead of $[t, t+\epsilon]$, we have, by definition of density, $d\nu(I_\epsilon) = \int_{I_\epsilon} h(s)\, d\mu(s)$. Hence,

$$\| \frac{d\nu(I_\epsilon)}{d\mu(I_\epsilon)} - h(t) \| \leq \| \frac{d\nu(I_\epsilon)}{d\mu(I_\epsilon)} - z_n \| + \eta \leq \frac{1}{d\mu(I_\epsilon)} \| \int_{I_\epsilon} [h(s) - z_n]\, d\mu(s) \| + \eta ,$$

and so

$$\| \frac{d\nu(I_\epsilon)}{d\mu(I_\epsilon)} - h(t) \| \leq \frac{d\nu_n(I_\epsilon)}{d\mu(I_\epsilon)} + \eta .$$

Letting $\epsilon\!\downarrow\!0$ and using (38), we obtain that the limsup of the left-hand side is less than or equal to $\| h(t) - z_n \| + \eta \leq 2\eta$. In view of η being arbitrary, this implies (39) and proves the result. $\qquad\qquad\square$

In [Mor-Val 1] is given a different proof, which does not depend on Jeffery's result. It proceeds by "unfolding the jumps" of a function of bounded variation and reducing the problem to the study of differential measures of Lipschitz functions (to which the results on Lebesgue points may be applied).

We must point out that the earlier "bilateral" version due to Daniell (1918), namely that

$$(40) \qquad \frac{d\nu}{d\mu}(t) = \lim_{\epsilon\!\downarrow\!0} \frac{d\nu([t-\epsilon, t+\epsilon])}{d\mu([t-\epsilon, t+\epsilon])} ,$$

sometimes is not enough for our study of differential inclusions. We shall often refer to Theorem 1.1 simply as Jeffery's theorem.

0.2. Compactness results for functions of bounded variation

We prove some properties, needed in the sequel, concerning extraction of convergent subsequences of functions of bounded variation.

Theorem 2.1. *Let (u_n) be a sequence of functions from the interval $I = [0, T]$ to a Hilbert space H. Assume that (u_n) is uniformly bounded in norm and in variation, i.e., that there exist $L, M > 0$ such that:*

$$(1) \qquad \| u_n(t) \| \leq L \qquad (t \varepsilon I,\ n \varepsilon \mathbb{N})$$

$$(2) \qquad \mathrm{var}\,(u_n,I) \leq M \qquad (n \varepsilon \mathbb{N})\,.$$

(**i**) *Then, there exists a subsequence (u_{n_k}) of (u_n) which converges pointwisely weakly to some function $u \colon I \to H$ with variation $\leq M$:*

$$(3) \qquad \text{w-}\lim_k\ u_{n_k}(t) = u(t) \qquad (t \varepsilon I)\,.$$

(**ii**) *Moreover, if the functions (u_n) and u are left-continuous, then for every continuous function or right-continuous function of bounded variation $\phi \colon I \to H$ we have:*

$$(4) \qquad \int_{[s,\,t[} \phi \cdot du_{n_k}\ \to\ \int_{[s,\,t[} \phi \cdot du \quad (s < t)\,.$$

(**iii**) *If the functions (u_n) and u are right-continuous, then for every continuous function or left-continuous function of bounded variation $\phi \colon I \to H$ we have:*

$$(5) \qquad \int_{]s,\,t]} \phi \cdot du_{n_k}\ \to\ \int_{]s,\,t]} \phi \cdot du \quad (s < t)\,.$$

Remark. If H is finite-dimensional, (3) is of course equivalent to strong convergence at every t:

$$(6) \qquad \| u_{n_k}(t) - u(t) \|\ \to\ 0 \qquad (t \varepsilon I).$$

In this case, the result is due to Helly ("the first Helly theorem") and to Banach ([Ban], p. 173-4), so that we may refer to this type of result as *Helly-Banach's theorem*. Castaing has given in [Cas 1] (Theorem 3) an analogous result for functions with values in a separable Hilbert space (without loss of generality) but extracting instead a generalized subsequence or subnet of the given sequence. In [Bar-Pre] is proved a generalization of (i) to functions taking values in Banach spaces.

Proof. (i) Every function u_n has bounded variation, hence its image is contained in a separable subspace of H. Then there exists a closed separable subspace, say H_0, which contains $u_n(t)$ for all n and all t. Considering an orthogonal decomposition of H, it is easily verified that we only need to prove the weak convergence in H_0. In other words, we may assume that the initial Hilbert space is *separable*.

Let (e_m) $(m \varepsilon \mathbb{N})$ be a complete orthonormal basis in H — only the infinite dimensional case, not treated in [Ban], interests us here. For every m, the real functions $u_n(t) \cdot e_m$ $(n \varepsilon \mathbb{N})$ are uniformly bounded by L and uniformly bounded in variation by M.

Banach's result, as mentioned above, implies that there is a subsequence $\mathcal{S}_1$ such that $u_n(t) \cdot e_1$ (for $u_n \varepsilon \mathcal{S}_1$) converges everywhere to a bv function $f_1(t)$; and then there exists a subsequence $\mathcal{S}_2$ of $\mathcal{S}_1$ such that $u_n(t) \cdot e_2$ converges pointwisely to a bv function $f_2(t)$ along $\mathcal{S}_2$; and so on. The classical diagonalization procedure furnishes in this manner a subsequence (u_{n_k}) of (u_n) for which

$$u_{n_k}(t) \cdot e_m \to f_m(t),$$

where (f_m) is a sequence of bv functions from I to $\mathbb{R}$.

For every t in I, $T(h) := \lim_k u_{n_k}(t) \cdot h$ defines a continuous linear functional on H, with $T(e_m) = f_m(t)$ and $|T(h)| \le L \|h\|$. By Riesz representation theorem, there is a unique $u(t) \varepsilon H$ such that $T(h) = u(t) \cdot h$ holds for every $h \varepsilon H$. Then (u_{n_k}) converges pointwisely weakly to this function $u: I \to H$. Moreover, u has bounded variation. In fact, given any choice of nodes in I, we have, by the lower semicontinuity of the norm in the weak topology of H:

$$\sum_{i=2}^{p} \| u(t_i) - u(t_{i-1}) \| \le \sum_{i=2}^{p} \underline{\lim} \| u_{n_k}(t_i) - u_{n_k}(t_{i-1}) \|$$

$$\le \underline{\lim} \sum_{i=2}^{p} \| u_{n_k}(t_i) - u_{n_k}(t_{i-1}) \| \le \underline{\lim} \, \mathrm{var}(u_{n_k}, I) \le M,$$

whence $\mathrm{var}(u; I) \le M$.

(ii) We assume that ϕ is continuous or that ϕ is right-continuous and of bounded variation. Given any $\epsilon > 0$ let us take a partition $J_i = [s_i, s_{i+1}[$ of $J = [s, t[$ (with $i = 0, \ldots, m$) such that the oscillation of ϕ in every subinterval is not greater than ϵ. The right-continuous step-function ψ, defined by $\psi(\tau) := \phi(s_i)$ if $\tau \varepsilon J_i$, satisfies $\| \psi - \phi \|_\infty \le \epsilon$ and

$$\int_J \psi \cdot du_{n_k} = \sum_{0 \le i \le m} \phi(s_i) \cdot \int_{[s_i, s_{i+1}[} du_{n_k} \cdot$$

Because the functions u_n and u are left-continuous and by (3), we have:

$$\lim_k \int_{[s_i, s_{i+1}[} du_{n_k} = \lim_k [u_{n_k}(s_{i+1}) - u_{n_k}(s_i)] = u(s_{i+1}) - u(s_i)$$

$$= \int_{[s_i, s_{i+1}[} du,$$

and so

$$\lim_k \int_J \psi \cdot du_{nk} = \sum_{0 \le i \le m} \phi(s_i) \cdot \int_{[s_i, s_{i+1}[} du = \int_J \psi \cdot du .$$

We take k_0 such that, for $k \ge k_0$, $\;| \int_J \psi \cdot du_{nk} - \int_J \psi \cdot du | \le \epsilon$.

On the other hand, we recall that by (2) and (i) the total variations of u_n and u (which equal the norms of the measures du_n and du, since u_n and u have aligned jumps; cf. (1.17)) are bounded above by M.

Hence, for $k \ge k_0$:

$$| \int_J \phi \cdot du_{nk} - \int_J \phi \cdot du | \le \int_J \| \phi - \psi \| \; | du_{nk} | + | \int_J \psi \cdot du_{nk} - \int_J \psi \cdot du |$$
$$+ \int_J \| \psi - \phi \| \; | du | \le (2M+1) \, \epsilon,$$

proving (4).

(iii) It is analogous to (ii). $\qquad\square$

Next, we consider generalized sequences or nets of functions of bounded variation. We obtain a similar result, with a small refinement.

Theorem 2.2. *Let (u_α) be a generalized sequence or net of functions of bounded variation from $I=[0,T]$ to a real Hilbert space H. Assume that (u_α) is uniformly bounded in norm and in variation:*

$$(7) \qquad\qquad \| u_\alpha \|_\infty \le L \; ; \; \mathrm{var} \, (u_\alpha \; ; \; I) \le M .$$

(i) There is a filter $\mathcal{F}$ finer than the filter of the sections of the index set (in other words, there is a subnet extracted from the given net) and there is a function of bounded variation $u: I \to H$ such that:

$$(8) \qquad\qquad \text{w-}\lim_{\mathcal{F}} u_\alpha(t) = u(t) ,$$

$$(9) \qquad\qquad \mathrm{var}(u;I) \le M.$$

(ii) Moreover, if all the functions u_α are right-continuous, then for every continuous $\phi: I \to H$ and $s<t$ in I we have:

$$(10) \;\; \lim_{\mathcal{F}} \int_{]s,t]} \phi \cdot du_\alpha = \int_{]s,t]} \phi \cdot du + \phi(s) \cdot [u^+(s) - u(s)] - \phi(t) \cdot [u^+(t) - u(t)].$$

In particular, if the (sub)limit function u is right-continuous at both s and t:

$$(11) \qquad\qquad \lim_{\mathcal{F}} \int_{]s,t]} \phi \; du_\alpha = \int_{]s,t]} \phi \; du .$$

Proof. (i) Since the net is uniformly bounded and balls are weakly compact in H, it follows from Tychonoff's theorem that (u_α) is relatively compact in the topology of pointwise weak convergence. Thus, there exist $\mathcal{F}$ and u satisfying (8). Inequality (9) is obtained as in the preceding theorem.

(ii) Being a function of bounded variation, u has a countable set of discontinuity points. Hence, for every n, there exist $t_{n,0} = s < t_{n,1} < \dots < t_{n,\nu} < t_{\nu+1} = t$ (ν varying with n) such that $t_{n,i+1} - t_{n,i} \leq 1/n$ and u is continuous at $t_{n,i}$ for $0 < i \leq \nu$.

Let us define $\phi_n(\tau) = \phi(t_{n,i+1})$ if $t\varepsilon\,]t_{n,i}, t_{n,i+1}]$ and $0 \leq i \leq \nu$. Since every u_α is right-continuous and by making use of (8):

$$\lim_{\mathcal{F}} \int_{]s,t]} \phi_n \cdot du_\alpha = \lim_{\mathcal{F}} \sum_{i=0}^{\nu} \phi(t_{n,i+1}) \cdot [u_\alpha(t_{n,i+1}) - u_\alpha(t_{n,i})]$$

$$= \sum_{i=0}^{\nu} \phi(t_{n,i+1}) \cdot [u(t_{n,i+1}) - u(t_{n,i})] .$$

On the other hand, we have:

$$\int_{]s,t]} \phi_n \cdot du = \sum_{i=0}^{\nu} \phi(t_{n,i+1}) \cdot [u^+(t_{n,i+1}) - u^+(t_{n,i})]$$

$$= \sum_{i=1}^{\nu-1} \phi(t_{n,i+1}) \cdot [u(t_{n,i+1}) - u(t_{n,1})] + \phi(t_{n,1}) \cdot [u(t_{n,1}) - u^+(s)]$$

$$+ \phi(t) \cdot [u^+(t) - u(t_{n,\nu})] .$$

Comparing these expressions:

$$(12) \qquad \lim_{\mathcal{F}} \int_{]s,t]} \phi_n \cdot du_\alpha = \int_{]s,t]} \phi_n \cdot du + \phi(t_{n,1}) \cdot [u^+(s) - u(s)]$$

$$+ \phi(t) \cdot [u(t) - u^+(t)].$$

The estimate

$$\left| \int_{]s,t]} \phi_n \cdot du_\alpha - \int_{]s,t]} \phi \cdot du_\alpha \right| \leq M \, \| \phi_n - \phi \|_\infty$$

and the uniform convergence of (ϕ_n) to ϕ imply that

$$(13) \qquad\qquad \lim_n \int_{]s,t]} \phi_n \cdot du_\alpha = \int_{]s,t]} \phi \cdot du_\alpha ,$$

uniformly with respect to α. Then, by Moore's theorem (see [Dun-Sch] I.7.6) we may change the order in the following double limit, using (12) and (13):

$$\lim_{\mathcal{F}} \int_{]s,t]} \phi \cdot du_\alpha = \lim_{\mathcal{F}} \lim_n \int_{]s,t]} \phi_n \cdot du_\alpha = \lim_n \lim_{\mathcal{F}} \int_{]s,t]} \phi_n \cdot du_\alpha$$

$$= \lim_n \left[\int_{]s,t]} \phi_n \cdot du + \phi(t_{n,1}) \cdot (u^+(s) - u(s)) - \phi(t) \cdot (u^+(t) - u(t)) \right].$$

To conclude it suffices to remark that $\phi(t_{n,1}) \to \phi(s)$, because ϕ is continuous and $|t_{n,1} - s| \leq 1/n$, and that $\int_{]s,t]} \phi_n \cdot du \to \int_{]s,t]} \phi \cdot du$ is obtained in the same way as (13). $\square$

We leave it to the reader to formulate and prove the analogue of (ii) for functions of bounded variation which are left-continuous.

0.3. Convergence in the sense of filled-in graphs

In this section, we introduce a notion of convergence which seems well-adapted to the study of discontinuous functions, by providing a reasonable substitute for uniform convergence. In view of the applications we have in mind, we shall restrict ourselves to considering right-continuous functions of bounded variation, *rcbv functions* for short.

Let $f: I \to H$ be an rcbv function from a real interval $I = [0, T]$ to a Hilbert space H. We define the *filled-in graph* $gr^* f$ by adding, if necessary, some line segments to the graph of f in such a way that all its gaps, if any, are filled. To be precise,

$$gr^* f = \{ (t, x) : \; 0 \leq t \leq T \text{ and } x \; \varepsilon \; [\, f^-(t), f(t)] \},$$

where $[y, z]$ stands for the line segment between two points in H and $f^-(t)$ is the left limit of f at point t. Here, by convention, $f^-(0) = f(0)$.

Filled-in graphs of such rcbv functions are closed bounded subsets of the product space $I \times H$; hence, we may consider the Hausdorff distance between any two of them with respect to a suitable metric, for instance,

$$\delta \left((t,x), (s,y) \right) = \max \{ \, |\, t - s\, | \, , \, \|\, x - y\, \| \, \}.$$

Recall that in a metric space (X, δ) the *Hausdorff distance* between two (non-empty closed) subsets A and B is defined by

$$h(A, B) = \max \{ \, e(A, B) \, , \, e(B, A) \, \},$$

where $e(A, B)$ is the excess or separation of A from B which is given by

$$e(A, B) = \sup \{ \, \text{dist}(a, B) : \, a \varepsilon A \} = \sup_{a \varepsilon A} \; \inf_{b \varepsilon B} \; \delta(a, b).$$

If f and g are two rcbv functions, we define

$$h^*(f, g) := h(gr^* f, gr^* g),$$

that is, the Hausdorff distance between their filled-in graphs with respect to the metric δ introduced above. This is a metric in $rcbv(\,I,\,H\,)$, because of the general properties of Hausdorff distances and of the following:

Proposition 3.1. *If f and g are rcbv functions such that $h^*(f,\,g) = 0$, then $f = g$.*

Proof. Suppose, by contradiction, that $f(t_0) \neq g(t_0)$ for some t_0 in $[0, T[$. Since both functions are right-continuous, there is $\epsilon > 0$ such that if both t and s belong to $J = \,]t_0,\, t_0 + \epsilon[$, then $\| f(t) - f(t_0) \| < \delta$ and $\| g(s) - g(t_0) \| < \delta$ where $\delta = \| f(t_0) - g(t_0) \| / 3$. It follows that $\| g^-(s) - g(t_0) \| \leq \delta$. Hence, if $(s, x) \in gr^* g$ with $s \in J$, i.e., $x \in [g^-(s),\, g(s)]$, then $\| x - g(t_0) \| \leq \delta$; so that $\| x - f(t) \| \geq \delta$, for every t in J. Let $t_1 = t_0 + \epsilon/2$. If (s, x) belongs to the filled-in graph of g, then either $s \notin J$ and so $|s - t_1| \geq \epsilon/2$; or $s \in J$ and then $\| x - f(t_1) \| \geq \delta$. We conclude that

$$h^*(f, g) \;\geq\; \mathrm{dist}\,((t_1,\, f(t_1))\,,\, gr^* g) \;\geq\; \min\,\{\,\epsilon/2\,,\,\delta\,\} > 0,$$

which contradicts the initial hypothesis.

We have shown so far that $f = g$ on $[0,\, T[$, which also implies $f^-(T) = g^-(T)$. Suppose that f is left-continuous at T. Since $\mathrm{dist}\,((T, g(T)),\, gr^* f) = 0$, by hypothesis, then there exist sequences $t_n \to T$ and $x_n \in [f^-(t_n),\, f(t_n)]$ with $x_n \to g(T)$. Both $f^-(t_n)$ and $f(t_n)$ converge to $f^-(T) = f(T)$, and this forces x_n to do the same. Hence $f(T) = g(T)$; in particular, g is also left-continuous at T. Finally, we suppose that f and g are both discontinuous at T. Let $x \in \,]f^-(T), f(T)]$ (in an obvious sense) and consider a sequence $(t_n,\, x_n)$ in $gr^* g$ which converges to (T, x). Then $t_n = T$ for all n large enough, otherwise x_n would converge to $f^-(T) \neq x$. Hence x_n belongs to $[g^-(T),\, g(T)]$ and so does the limit x. From the previous considerations and exchanging the roles of f and g we infer that line segments $[f^-(T), f(T)]$ and $[g^-(T),\, g(T)] = [f^-(T), g(T)]$ are the same, whence $f(T) = g(T)$. $\qquad\qquad\square$

The next simple result relates convergence in the sense of filled-in graphs with uniform convergence.

Proposition 3.2. *If f and g are rcbv functions, then*

$$(1) \qquad\qquad\qquad h^*(f, g) \;\leq\; \| f - g \|_\infty\,.$$

Thus, every sequence of rcbv functions which converges uniformly to an rcbv function also converges in the sense of filled-in graphs to the same function.

Proof. Let $r = \| f - g \|_\infty$, the sup norm. For every t in I, $\| f(t) - g(t) \| \le r$. Since the same is true for every $s < t$, we have $\| f^-(t) - g(t) \| \le r$. Hence:

$$h\big(\, [f^-(t), f(t)] \, , \, [g^-(t), \, g(t)] \, \big) \; \le \; r \, ,$$

with respect to the norm metric of H and (1) follows easily. □

This notion of convergence of filled-in graphs presents an inconvenient feature, already detected for the convergence of graphs, as introduced by Moreau [Mor 4]: a sequence of filled-in graphs may converge in $I \times H$, in the sense of Hausdorff distance, to a limit set L without L being the filled-in graph of an rcbv function.

The following example shows this and also makes it clear that, in general, from a sequence of rcbv functions, which are uniformly bounded both in norm and in variation, it is not possible to extract a subsequence which converges in the sense of filled-in graphs to an rcbv (or simply bv) function, even if the functions take their values in a finite dimensional space.

Example 3.3. Consider the functions from $[0, 1]$ to $\mathbb{R}$ defined by

$$u_n(t) = \begin{cases} -1, & \text{if } 1/3n \le t < 1/2n \\ 1, & \text{if } 1/2n \le t < 1/n \\ 0, & \text{otherwise.} \end{cases}$$

We have $\| u_n(t) \| \le 1$, $\operatorname{var}(u_n; 0, 1) = 4$ $(n \ge 2)$ and (u_n) converges pointwisely to the zero function. With respect to Hausdorff distance in $\mathbb{R}^2$, their filled-in graphs clearly converge to the set

$$L = \{0\} \times [-1, 1] \; \cup \;]0, 1] \times \{0\},$$

which is not a filled-in graph.

Under stronger hypotheses the extraction of convergent subsequences becomes possible, namely if the measures of total variation of the given functions, denoted by $| du_n |$, are all bounded above – in the sense of the ordering of real measures – by the same positive bounded measure.

Theorem 3.4. *Consider a sequence (u_n) of rcbv functions from $I = [0, T]$ to an Euclidian space E. Suppose that there exist a positive constant L and a finite measure $d\nu \ge 0$ such that for every n:*

$$(2) \hspace{4cm} \| u_n(0) \| \le L \, ,$$

$$(3) \qquad\qquad\qquad | \, du_n \, | \; \leq \; d\nu \; .$$

Then, we can extract a subsequence (u_{n_k}) which converges in the sense of filled-in graphs to an rcbv function u that satisfies $| \, du \, | \leq d\nu$.

Notice that $\mathrm{var}(u_n \, , I) \leq \| \, d\nu \, \| = \int d\nu$ and $\| \, u_n(t) \, \| \leq L + \| \, d\nu \, \|$. Thus, by Theorem 2.1 there exists at least a subsequence which converges pointwisely to a bv function u. Thanks to (3), we can prove more:

Lemma 3.5. *Under the hypotheses of Theorem 3.4, there exist a subsequence (u_{n_k}) and an rcbv function u such that $| \, du \, | \leq d\nu$ and for every $t \varepsilon I$:*

$$(4) \qquad\qquad\qquad u_{n_k}(t) \to u(t)$$

$$(5) \qquad\qquad\qquad u_{n_k}^{\;-}(t) \to u^-(t) \; .$$

Proof. By (3), every Stieltjes measure du_n is absolutely continuous with respect to the measure $d\nu$. Hence, du_n has a density $\gamma_n = \dfrac{du_n}{d\nu}$ which belongs to the unit ball of $L_E^\infty = L_E^\infty(I, \, d\nu; \, E)$. This ball is a weak-$*$ compact subset of L_E^∞, so that by Eberlein-Smulian theorem it is also sequentially $\sigma(L_E^1, \, L_E^\infty)$-compact. Then there exists a subsequence (γ_{n_k}) which converges weakly to some γ in L_E^1 ; moreover, $\| \, \gamma(t) \, \| \leq 1$ $d\nu$-almost everywhere in I. By (2), we may suppose additionally that $u_{n_k}(0)$ converges to some $a \, \varepsilon \, E$.

If $x \varepsilon E$ and $t \varepsilon I$, then, because we are dealing with right-continuous functions:

$$x \, . \, u_{n_k}(t) = x \, . \, u_{n_k}(0) + \int_{]0, \, t]} x \, . \, du_{n_k}(s) = x \, . \, u_{n_k}(0) + \int_{]0, \, t]} x \, . \, \gamma_{n_k}(s) \; d\nu(s).$$

We denote by $< \, . \, , \, . \, >$ the duality product between L_E^1 and L_E^∞ and by χ_A the characteristic function of a subset of I. We have:

$$\lim_k \, x \, . \, u_{n_k}(t) = \lim_k \, [\; x \, . \, u_{n_k}(0) + < \, \gamma_{n_k} \, , \; \chi_{]0, \, t]} \, x \; > \;]$$

$$= x \, . \, a + \; < \, \gamma \, , \; \chi_{]0, \, t]} \, x \; > = x \, . \, a + \int_{]0, \, t]} x \, . \, \gamma(s) \; d\nu(s).$$

Thus, $u_{n_k}(t)$ converges weakly and strongly, since E is finite-dimensional, to

$$(6) \qquad\qquad u(t) = a + \int_{]0, \, t]} \gamma(s) \; d\nu(s).$$

The function u defined in this manner is clearly rcbv and

$$(7) \qquad\qquad | \, du \, | = | \, \gamma \, d\nu \, | = | \, \gamma \, | \; d\nu \leq d\nu.$$

Finally, (5) is obtained in the same way as (4):

$$\lim_k \, x \cdot u_{nk}^{-}(t) = \lim_k \, [\, x \cdot u_{nk}(0) + \int_{]0,\,t[} x \cdot \gamma_{nk}(s) \; d\nu(s)\,]$$

$$= x \cdot [\, a + \int_{]0,\,t[} \gamma(s) \; d\nu(s)\,] = x \cdot u^{-}(t) \; . \qquad \square$$

This section ends with the:

Proof of Theorem 3.4. Let $\epsilon > 0$ be given. We choose $t_0 < t_1 < \ldots < t_m = T$ such that for every i:

$$(8) \qquad\qquad |\, t_{i+1} - t_i \,| \leq \epsilon \,,$$

$$(9) \qquad\qquad d\nu \, (\,]\, t_i \,, t_{i+1}[\,) \leq \epsilon \,.$$

Since $\phi(t) := d\nu\,([0,t])$ is a bounded nondecreasing function, then only a finite number of the intervals $J_k := \phi^{-1}(\,[\,k\epsilon,(k+1)\epsilon[\,)$, with $k = 0, 1, \ldots$ are nonempty. We take their endpoints and add the intermediate points needed to satisfy (8). With this construction, if s and t belong to the same $I_i = \,]\,t_i\,,t_{i+1}[$ with $s < t$, then they also are in the same J_k and so

$$d\nu\,(\,]s,t]) = d\nu\,([0,t]) - d\nu\,([0,s]) \leq (k+1)\epsilon - k\epsilon = \epsilon \; ;$$

now (9) follows by taking the supremum with respect to s and t.

From (3), (7) and (9) we derive estimates on the oscillations:

$$(10) \qquad\qquad \mathrm{osc}\,(u_n;\, I_i) \leq \epsilon \,, \; \mathrm{osc}\,(\,u;\, I_i) \leq \epsilon \,.$$

By Lemma 3.5, there is an integer N such that for $n \geq N$ we have, for every i:

$$(11) \qquad\qquad \|\, u_n\,(t_i) - u\,(t_i)\,\| \leq \epsilon \,,$$

$$(12) \qquad\qquad \|\, u_n^{-}(t_i) - u^{-}(t_i)\,\| \leq \epsilon \,.$$

The proof will be complete if we show that:

$$(13) \qquad\qquad h(\, gr^{*}\, u_n \,, \, gr^{*}\, u \,) \leq 2\epsilon \,, \; \text{whenever} \; n \geq N .$$

First, we prove that:

$$(14) \qquad\qquad e(\, gr^{*}\, u_n \,, \, gr^{*}\, u \,) \leq 2\epsilon .$$

If (t, x) is in the filled-in graph of u_n, i. e., $x \varepsilon\,[u_n^{-}(t),\, u_n(t)]$ and if $t \varepsilon I_i$, then $|\, t - t_i \,| \leq \epsilon$, by (8). Right-continuity of u_n, (10) and (11) imply that

$$\|\, u_n(t) - u(t_i)\,\| \leq \|\, u_n^{+}(t) - u_n^{+}(t_i)\,\| + \|\, u_n(t_i) - u(t_i)\,\|$$

$$\leq \mathrm{osc}(u_n \,,\, I_i) + \epsilon \leq 2\epsilon \,,$$

and analogously $\| u_n^-(t) - u(t_i) \| \leq 2\epsilon$. This implies $\| x - u(t_i) \| \leq 2\epsilon$ and so

$$\mathrm{dist}((t,x), gr^*u) \leq \delta((t,x),(t_i, u(t_i))) \leq \max\{\epsilon, 2\epsilon\} = 2\epsilon .$$

If t is one of the points t_i, then x belongs to the segment $[u_n^-(t_i), u_n(t_i)]$ whose Hausdorff distance in E to $[u^-(t_i), u(t_i)]$ is not greater than ϵ, by (11) and (12). It follows easily that

$$\mathrm{dist}((t_i, x), gr^*u) \leq \mathrm{dist}((t_i, x), \{t_i\} \times [u^-(t_i), u(t_i)]) \leq \epsilon ,$$

thus ending the proof of (14).

We conclude by proving that, symetrically:

$$(15) \qquad\qquad e(gr^* u , gr^* u_n) \leq 2\epsilon .$$

If $(t_i, x) \epsilon\, gr^* u$, then as above:

$$\mathrm{dist}((t_i, x), gr^*u_n) \leq \mathrm{dist}((t_i, x), \{t_i\} \times [u_n^-(t_i), u_n(t_i)])$$

$$\leq h([u^-(t_i), u(t_i)], [u_n^-(t_i), u_n(t_i)]) \leq \epsilon .$$

If $t\epsilon\, I_i$ and $x \,\epsilon\, [u-(t), u(t)]$, then $|t - t_i| < \epsilon$ and

$$\| u-(t)-u_n(t_i) \| \leq \| u-(t)-u^+(t_i) \| + \| u(t_i)-u_n(t_i) \| \leq \mathrm{osc}(u; I) + \epsilon \leq 2\epsilon .$$

Similarly, $\| u(t) - u_n(t_i) \| \leq 2\epsilon$.

Hence $\| x - u_n(t_i) \| \leq 2\epsilon$ and $\mathrm{dist}((t, x), gr^*u_n) \leq 2\epsilon$. $\qquad\square$

0.4. Geometrical inequalities

The first inequality presented here concerns the Hausdorff distance h between subsets of a Hilbert space H. It implies that if a continuous rectifiable arc l has length not much greater than that of a certain line segment and if the respective endpoints are not very far apart, then the Hausdorff distance between arc and segment is small. We do not aim at giving the optimal quantification of this property.

Lemma 4.1. *Let $[x_1, x_2]$ be a line segment in H. Let l be a continuous rectifiable arc with length $|l|$ not greater than $\| x_1-x_2 \| + \nu$ and with endpoints y_1 and y_2 such that $\| y_1 - x_1 \| \leq \delta$ and $\| y_2 - x_2 \| \leq \eta$. Here ν, δ and η are nonnegative numbers such that $\nu+\delta+\eta \leq 1$. Then*

$$(1) \qquad h(l, [x_1, x_2]) \leq (3 + \| x_1 - x_2 \|)\sqrt{\nu+\delta+\eta} .$$

Proof. First we assume that $\delta = \eta = 0$, that is, the endpoints of the arc and the segment are the same, and we show that, for every z in the arc l, the following estimate holds:

$$(2) \qquad \operatorname{dist}(z, [x_1, x_2]) \leq (1 + \| x_1 - x_2 \|) \sqrt{\nu} \, .$$

Hereafter, we shall denote the segment by S and its length by $|S|$. If (2) is not satisfied at some point $z' \varepsilon \, l$ and if x_1 is the proximal point of z' in S, we immediately see that $\| z' - x_2 \| \geq |S|$ and

$$| l | \geq \| x_1 - z' \| + \| z' - x_2 \| > (1 + |S|) \sqrt{\nu} + |S| \, .$$

Since $0 \leq \nu \leq 1$, this implies $|l| > \nu + |S|$, a contradiction. If the nearest point is x_2, we use a similar argument. If neither is the case, let x' be the orthogonal projection (nearest point) of z' in S. Observing that for $i = 1, 2$

$$\| z' - x' \|^2 > (1 + |S|)^2 \, \nu \geq \nu + 2 \, |S| \, \nu \geq \nu^2 + 2\nu \| x_i - x' \| \, , \text{ we write:}$$

$$| l | \geq \sum_{i=1}^{2} \| x_i - z' \| \geq \sum_{i=1}^{2} \sqrt{\| x_i - x' \|^2 + \| x' - z' \|^2}$$

$$\geq \sum_{i=1}^{2} (\| x_i - x' \| + \nu) = \| x_1 - x_2 \| + 2\nu \, ,$$

again contradicting the hypothesis. So, (2) is proved.

Reciprocally, notice that every x' in S is the orthogonal projection of some z' in the arc l. In fact, these projections are the image of an interval by a continuous function; thus, they form a connected subset of the line segment containing its endpoints, hence they equal the line segment itself. Therefore,

$$\operatorname{dist}(x', l) \leq \| x' - z' \| = \operatorname{dist}(z', S) \leq (1 + |S|) \sqrt{\nu} \, .$$

This proves that, if the arc l has endpoints x_1 and x_2 and length $|l| \leq |S| + \nu \ (0 \leq \nu \leq 1)$, then:

$$(3) \qquad h(l, S) \leq (1 + |S|) \sqrt{\nu} \, .$$

In the general case, we remark that $| \, \| x_1 - x_2 \| - \| y_1 - y_2 \| \, | \leq \delta + \eta \leq 1$, so that $|l| \leq \| y_1 - y_2 \| + \delta + \eta + \nu$. Thus, we may apply (3) to l and to the segment $S' = [y_1, y_2]$:

$$h(l, S') \leq (1 + |S'|) \sqrt{\nu + \delta + \eta} \leq (2 + |S|) \sqrt{\nu + \delta + \eta} \, .$$

To obtain (1), we just use the triangular inequality for the Hausdorff distance and the estimate:

$$h(S', S) \leq \max \| x_i - y_i \| \leq \delta + \eta \leq \sqrt{\nu + \delta + \eta} \, . \qquad \square$$

The next set of results, by Moreau and Valadier, is essential in order to estimate the variation of the solutions to a new class of sweeping processes.

Lemma 4.2. [Mor 8] *Let S be a contraction from a subset D of a Hilbert space into itself. Assume that P, the set of fixed points of S, contains some closed ball $\overline{B}(a, r)$, $r > 0$. Then, for every $x \varepsilon D$, we have:*

$$(4) \qquad \| x - S(x) \| \leq \frac{1}{2r} (\| x - a \|^2 - \| S(x) - a \|^2) .$$

Proof. Let y be such that $\| y \| = 1$ and $x - S(x) = \| x - S(x) \| \, y$. Since $a + ry \varepsilon P$ and since S is a contraction, it follows that $\| a + ry - x \|^2 \geq \| a + ry - S(x) \|^2$, i. e.,

$$r^2 + 2\, ry . (a - x) + \| a - x \|^2 \geq r^2 + 2\, r y . (a - S(x)) + \| a - S(x) \|^2 .$$

Then,

$$\frac{1}{2r} (\| x - a \|^2 - \| S(x) - a \|^2) \geq y . (x - S(x)) = \| x - S(x) \| . \qquad \square$$

An easy consequence of Lemma 4.2 is:

Lemma 4.3. *In a Hilbert space H, we consider a closed convex set C which contains a closed ball $\overline{B}(a, r)$ $(r > 0)$. Let $x \varepsilon H$. Then:*

$$(5) \qquad \| x - \operatorname{proj}(x, C) \| \leq \frac{1}{2r} (\| x - a \|^2 - \| \operatorname{proj}(x, C) - a \|^2) .$$

Proof. [Val 2] It suffices to apply the preceding lemma to the projection of best approximation on C, $S(x) := \operatorname{proj}(x, C)$ (see p. 26) . In fact, S is a contraction of H having C as set of fixed points. $\qquad \square$

For iterative procedures we may use the following estimates.

Lemma 4.4. *Let $C_1, \ldots, C_n$ be closed convex subsets of H all of them containing a fixed closed ball $\overline{B}(a, r)$ $(r > 0)$. If $x_0 \varepsilon H$ and if $x_1, \ldots, x_n$ are defined inductively by $x_i = \operatorname{proj}(x_{i-1}, C_i)$, then:*

$$(6) \qquad \| x_0 - a \| \geq \| x_1 - a \| \geq \ldots \geq \| x_n - a \| ;$$

$$(7) \quad \sum_{i=1}^{n} \| x_i - x_{i-1} \| \leq \frac{1}{2r} (\| x_0 - a \|^2 - \| x_n - a \|^2) \leq \frac{1}{2r} \| x_0 - a \|^2 .$$

Proof. [Val 2] Inequalities (6), which hold even if $r = 0$, result from the fact that projections are contractions:

$$\| x_1 - a \| = \| \operatorname{proj}(x_0, C_1) - \operatorname{proj}(a, C_1) \| \leq \| x_0 - a \|$$

and so on.

Since $C_i \supset \bar{B}(a, r)$, then by the preceding lemma we know that

$$\| x_i - x_{i-1} \| \leq \tfrac{1}{2r} (\| x_{i-1} - a \|^2 - \| x_i - a \|^2)$$

for $i = 1, \ldots, n$; adding up these inequalities we obtain (7). $\qquad\square$

Remarks.

1) To be more precise, we can prove that ([Val 2])

$$(8) \qquad \sum_{i=1}^{n} \| x_i - x_{i-1} \| \leq \max \left\{ 0, \tfrac{1}{2r} (\| x_0 - a \|^2 - r^2) \right\}.$$

In fact, either $\| x_0 - a \| \leq r$ and then $x_i = x_0$ for every i, so that (8) is trivial; or $\| x_0 - a \| > r$ and it is easily verified that $\| x_n - a \| \geq r$, hence (8) follows from (7).

Notice that

$$\ell(r, \| x_0 - a \|) := \tfrac{1}{2r} (\| x_0 - a \|^2 - r^2)$$

is the length of the arc of the "développante" of the circle with radius r (centered at a) lying between this circle and the circle of radius $\| x_0 - a \|$ (this is a curve drawn by the end of a rigid bar rolling without sliding on a circle with radius r). It was under this form that Valadier initially stated the property (see [Cas 1] appendix and [Val 1]). His first proof, although more complicated, is nevertheless still interesting because of the underlying intuitive geometrical considerations. If $H = \mathbb{R}$, a better estimate is given in [Val 2].

2) This type of inequality had already been used by Moreau in [Mor 9] ((5.2), p. 31) and [Mor 8] (remark 2, p. 144). In the former, the sweeping process by a convex set having bounded but not necessarily continuous variation and containing the ball $\bar{B}(a, r)$ is considered; it is shown that the total variation of the solution is bounded above by $\| x_0 - a \|^2 / r$, where x_0 is the initial value. In the latter, which concerns the sweeping process by a convex set with absolutely continuous variation, it is proved that the absolutely continuous solution u satisfies a. e.

$$\| \tfrac{du}{dt} \| < - \tfrac{1}{2r} \tfrac{d}{dt} \| a - u \|^2 \, ;$$

by integration, this yields an estimate equivalent to (7):

$$\mathrm{var}(u; 0, T) \le \tfrac{1}{2r} \left(\| a - u(0) \|^2 - \| a - u(T) \|^2 \right) .$$

The next results deal with distances to complements or to intersections of convex sets. The proofs are reformulations of the original ones.

Proposition 4.5. [Mor 7]((6), p.173) *If B_1 and B_2 are two convex subsets of a Hilbert space H, with* $\mathrm{int}\, B_2 \ne \emptyset$, *then:*

$$(9) \qquad\qquad \forall a \varepsilon H: \quad \mathrm{dist}\,(a, H \backslash B_1) \le \mathrm{dist}\,(a, H \backslash B_2) + e(B_1, B_2) .$$

Proof. We may assume that $a = 0$ (by translation) and, to avoid (9) being trivial, that $\rho := \mathrm{dist}(0, H \backslash B_1) > 0$, that $r := e(B_1, B_2) < +\infty$ and that $\rho - r > 0$. We only need to show that

$$(10) \qquad\qquad B(0, \rho - r) \subset \overline{B}_2 .$$

In fact, the closure of B_2 coincides with the closure of its nonempty interior; hence $B(0, \rho - r)$ is contained in B_2 and this implies $\mathrm{dist}(0, H \backslash B_2) \ge \rho - r$.

Suppose that (10) is false: there is $x \notin \overline{B}_2$ with $\| x \| < \rho - r$. Then, by the well-known separation theorem (of a point and a closed convex set) there is $y \varepsilon H$ with $\| y \| = 1$ such that:

$$x . y > \alpha > b . y , \quad \forall b \varepsilon B_2 .$$

Take $z = z(\lambda) := x + (\lambda - \| x \|) y$, with $\lambda < \rho$ and $\lambda \to \rho$. Since $z \varepsilon B(0, \rho) \subset B_1$, we expect that $\mathrm{dist}\,(z, B_2) \le r$. But, if $\| h \| \le r + x . y - \alpha$, we have

$$(z + h) . y \ge x . y + (\lambda - \| x \|) - \| h \| \ge \lambda - \| x \| - r + \alpha > \alpha ,$$

for λ close enough to ρ (depending only on x); thus $z + h \notin B_2$ for any such h. Therefore, the distance of z to B_2 is not less than $r + x . y - \alpha > r$, a contradiction. $\qquad\qquad\square$

Proposition 4.6. [Mor 7]((12)) *Let A and B be two convex subsets of a Hilbert space H. If $a \varepsilon A$ and $B(a, \rho) \subset B$, then:*

$$(11) \qquad \forall x \varepsilon H: \ \mathrm{dist}(x, A \cap B) \le \left(1 + \frac{\| x - a \|}{\rho}\right) [\, \mathrm{dist}(x, A) + \mathrm{dist}(x, B) \,] .$$

Proof. We assume that $a = 0$, by translation. Also, since the hypotheses imply that the closure of $A \cap B$ equals $\overline{A} \cap \overline{B}$, we may assume that A and B are both closed. Let $y = \mathrm{proj}\,(x, A \cap B)$. Then $x - y \ne 0$ (otherwise (11) is trivial) and it belongs to the outward normal cone to $A \cap B$ at y. Under our assumptions on a, this cone is the sum of the two normal cones to A and B. Usual notations for the outward normal cone are $N_A(y)$ or $\partial\psi_A(y)$, the subdifferential of the indicator function $\psi_A(x) = 0$, if $x \varepsilon A$ and $+\infty$ otherwise. A well-known result from Convex Analysis implies that:

$$\partial\psi_{A \cap B} = \partial(\psi_A + \psi_B) = \partial\psi_A + \partial\psi_B ,$$

since both indicator functions are finite at a, one being continuous. Hence,

$$x - y = p + q \ , \ p \varepsilon N_A(y) \ , \ q \varepsilon N_B(y) \ .$$

If $p = 0$, then $y = \operatorname{proj}(x, B)$, $\operatorname{dist}(x, A \cap B) = \operatorname{dist}(x, B)$ and (11) is trivial. Similar remark if $q = 0$. Take $u = \| x - y \|^{-1} p$ and $v = \| x - y \|^{-1} q$ so that $x - y = \| x - y \| (u + v)$, $\| u + v \| = 1$. Since $u \varepsilon N_A(y)$ and $0 \varepsilon A$, we have $u . (y - 0) \geq 0$; and $v \varepsilon N_B(y)$, $B(0, \rho) \subset B$ imply that

$$v . (y - \rho \frac{v}{\| v \|}) \geq 0.$$

Hence:

(12) $$u . y \geq 0 \ , \ v . y \geq \rho \| v \| \ .$$

We show that:

(13) $$(x - y) . u \leq \| u \| \operatorname{dist}(x, A) \ .$$

In fact, A is contained in the half-space $H(A) = \{ z : (z - y) . u \leq 0 \}$ and the projection of x in $H(A)$ is $z := x - \lambda u$, where $\lambda := \| u \|^{-2} (x - y) . u$ is positive (otherwise (13) is trivial). Hence, $\operatorname{dist}(x, A) \geq \| x - z \|$ and (13) follows, because $\| x - z \| = \lambda \| u \| = \| u \|^{-1} (x - y) . u$.

Similarly, $(x - y) . v \leq \| v \| \operatorname{dist}(x, B)$. Then

$$\operatorname{dist}(x, A \cap B) = \| x - y \| = (x - y) . (u + v)$$

implies

(14) $$\operatorname{dist}(x, A \cap B) \leq \| u \| \operatorname{dist}(x, A) + \| v \| \operatorname{dist}(x, B) ,$$

and also, in view of (12),

$$0 \leq \operatorname{dist}(x, A \cap B) \leq x . (u + v) - y . v \leq \| x \| - \rho \| v \| .$$

Thus:

$$\| v \| \leq \frac{\| x \|}{\rho} \ , \ \| u \| \leq 1 + \| v \| \leq 1 + \frac{\| x \|}{\rho} \ ,$$

so that the result follows from (14).   □

The last inequalities presented here concern projections or proximal points of two (different) points into two (different) convex subsets of a Hilbert space. Recall that $y = \text{proj}(x, C)$ if $y \varepsilon C$ and for every $z \varepsilon C$ $\|x - y\| \leq \|x - z\|$. Equivalently, y is also characterized by the *inequality of projections*:

$$\forall z \varepsilon C: \ (x - y).(y - z) \geq 0 \ ,$$

which also means that

$$x - y \varepsilon N_C(y) \ .$$

Proposition 4.7. [Mor 1] ((2.17)) *Let C, C' be two nonempty closed convex subsets of H and x, x' be two elements of H. Then:*

$$(15) \qquad \|\text{proj}(x, C) - \text{proj}(x', C')\|^2 \leq \|x - x'\|^2 + 2 \, \text{dist}(x, C) \, e(C', C)$$
$$+ 2 \, \text{dist}(x', C') \, e(C, C').$$

Hence,

$$(16) \qquad \|\text{proj}(x, C) - \text{proj}(x', C')\|^2 \leq \|x - x'\|^2$$
$$+ 2 \, h(C, C') \, [\, \text{dist}(x, C) + \text{dist}(x', C')\,]$$

and, if $y = \text{proj}(x, C)$ and $y' = \text{proj}(x', C')$, these are written as:

$$(17) \quad \|y - y'\|^2 - \|x - x'\|^2 \leq 2 \, \|x - y\| \, e(C', C) + 2 \, \|x' - y'\| \, e(C, C')$$
$$\leq 2 \, h(C, C') \, [\, \|x - y\| + \|x' - y'\|\,] \ .$$

Proof. Since, for any $u, v \varepsilon H$, $\|u\|^2 - \|v\|^2 \leq 2 \, u.(u - v)$, we have:

$$(18) \qquad \|y - y'\|^2 - \|x - x'\|^2 \leq 2 \, (y - y').(y - y' - x + x')$$
$$= 2 \, (x - y).(y' - y) + 2 \, (x' - y').(y - y') \ .$$

But, writing $z = \text{proj}(y', C) \varepsilon C$, we also know that $(x - y).(y - z) \geq 0$, because y is the proximal point of x in C. Hence,

$$(x - y).(y' - y) \leq (x - y).(y' - z) \leq \|x - y\| \, \text{dist}(y', C) \leq \|x - y\| \, e(C', C) \ .$$

Similarly, $(x' - y').(y - y') \leq \|x' - y'\| \, e(C, C')$, so that (18) implies (17). $\qquad\qquad \square$

Chapter 1

Regularization and Graph Approximation of a Discontinuous Evolution

1.1. Introduction

Moreau has shown that a Yosida-type regularization procedure can be used in order to prove the existence of a solution to the so-called sweeping process or evolution problem associated with a moving convex set (see e.g. [Mor 3]). If this set is supposed to be Lipschitz-continuous in the sense of Hausdorff distance, then the solution to the problem it defines is also Lipschitz-continuous with respect to a real variable t. The Yosida or Moreau-Yosida approximants, that is, the absolutely continuous solutions to the regularized differential equations derived from the initial problem, then converge uniformly to the solution to the sweeping process.

If we try to carry out a similar program for a discontinuous evolution, say one that is determined by a convex set with bounded variation, it is clear that uniform convergence is not to be expected since the limit solution is not necessarily a continuous function. Instead, a reasonable substitute is found in the concept of graph convergence, to be precise, convergence of filled-in graphs with respect to Hausdorff distance ($\S 0.3$). In this way, we account for uncertainties both in the values and in the arguments of the concerned functions, a feature that will certainly appeal to the numerical analyst.

We present here the general result concerning evolutions in arbitrary Hilbert spaces, which appeared in [Mon 5] — a previous preprint version dealt only with the finite-dimensional case. Notice the additional compactness hypothesis on the moving convex set, not needed in Moreau's theory (even for Yosida approximation of an absolutely continuous evolution, [Mor 3], $\S 5.g$). Example 1.3 below also shows that we cannot extend the results to multifunctions with right-continuous retraction.

Let H be a real Hilbert space with scalar product denoted by a dot and I be a compact interval $[0, T]$ of the real line. We consider a multifunction (that is, a set-valued function) $C : t \rightarrow C(t)$ from I to nonempty closed convex subsets of H. This multifunction, also referred as $C(.)$ or C, can be interpreted as describing the evolution of a convex "mobile" (moving set) between the instants $t = 0$ and $t = T$. We suppose that C is of bounded variation and right-continuous, rcbv for short; this means that its variation function $v(.)$ is right-continuous and finite on I. For our purposes, it is enough to recall that $v(t) := \operatorname{var}(C; 0, t)$ is nondecreasing and satisfies

$$(1) \qquad h(\, C(s), C(t)\,) \le v(t) - v(s) \quad (0 \le s \le t)\,,$$

where the h refers to Hausdorff distance between subsets of H, with respect to the norm metric.

Definition 1.1. A right-continuous function of bounded variation $w: I \to H$ is a **solution to the sweeping process** by the convex moving set $C(.)$ with initial value $a \varepsilon\, C(0)$ if it satisfies the following conditions:

$$(2) \qquad w(0) = a\,;$$

$$(3) \qquad w(t)\ \varepsilon\ C(t) \quad (t\varepsilon I)\,;$$

and there exists a positive measure $d\mu$ on I relative to which the Stieltjes measure dw of w has a density $w' \varepsilon\, L^1(I, d\mu; H)$, i. e., $dw = w'\, d\mu$, such that :

$$(4) \qquad -w'(t)\varepsilon\ N_{C(t)}\,(w(t))\,, \text{ for } d\mu\text{-almost every } t \text{ in } I.$$

This is written shortly as:

$$(5) \qquad -dw\ \varepsilon\ N_{C(t)}\,(w(t))\,.$$

We propose to approximate the solution to this differential inclusion, which is known to exist and to be unique [Mor 1], by solutions to suitable differential equations.

Definition 1.2. Given $\lambda > 0$, the respective **Yosida approximant** is the unique absolutely continuous solution to the Cauchy problem:

$$(6) \qquad \frac{du_\lambda}{dt}\,(t) + \tfrac{1}{\lambda}\,[\,u_\lambda(t) - \operatorname{proj}\,(\,u_\lambda(t), C(t))\,] \;=\; 0\,;$$

$$(7) \qquad u_\lambda(0) = a\,,$$

where (6) is to be satisfied by almost every t in I, in the sense of Lebesgue measure. In (6), $\operatorname{proj}(x, C(t))$ denotes the point of $C(t)$ which is closest to x, usually called the *projection* or the *proximal point* of x in $C(t)$.

As we have remarked already, in general the Yosida approximants do not converge uniformly to the solution w of the sweeping process, when $\lambda \to 0$; in fact, they are continuous and w is possibly discontinuous. We illustrate this with two simple examples. First, we recall that a function r is a retraction function, or better said a "super-retraction" of a multifunction C if the excess of $C(s)$ with respect to $C(t)$ satisfies

$$e(C(s), C(t)) \leq r(t) - r(s) , \quad \forall s \leq t .$$

Example 1.3. Let $H = \mathbb{R}$ and define C on the interval $I = [0,2]$:

$$C(t) = [0,2] \text{ if } t \neq 1; \quad C(1) = [1,2] .$$

Then C has right-continuous retraction: $r(t) = 0$, for $t < 1$ and $r(t) = 1$ for $t \geq 1$. Take $a = 0$ as initial value. It is easily seen that $u_\lambda \equiv 0$ for any $\lambda > 0$, while the solution to the sweeping process is $w(t) = 0$, if $t < 1$, and $w(t) = 1$, if $t \geq 1$. In this case, there is no convergence of (u_λ) to w in any reasonable (Hausdorff) topology.

Example 1.4. The preceding example is modified to ensure the right-continuity of the variation (and not merely of the retraction):

$$C(t) = [0,2] \text{ if } t < 1 ; \quad C(t) = [1,2] \text{ if } t \geq 1.$$

The solution to the sweeping process w is the same as before but the Yosida approximants are now given by:

$$u_\lambda(t) = \begin{cases} 0 & \text{if } 0 \leq t \leq 1 \\ 1 - e^{-(t-1)/\lambda} & \text{if } 1 < t < 2 \end{cases} .$$

When $\lambda \to 0^+$, $u_\lambda(t)$ converges pointwisely to the function u defined by $u(t) = 0$ if $t \leq 1$ and $u(t) = 1$ otherwise. We notice that w is the right-continuous regularization of u: $w = u^+$. Moreover, the graphs of the u_λ's converge in the sense of Hausdorff distance to the filled-in graph of w, which is obtained by adding the vertical segment between $(1,0)$ and $(1,1)$ to the graph of w.

These properties are also present in the general case, as stated in the main theorem:

Theorem 1.5 [Mon 5] *Let $C(.)$ be an rcbv (right-continuous, bounded variation) multifunction from $I = [0, T]$ to a real Hilbert space H, with nonempty closed convex values, and whose variation $v(.)$ is continuous at the right endpoint T.*

Moreover, suppose that, denoting by B the open unit ball of H:

(8) $C(t) \cap r\bar{B}$ *is strongly compact, for every $t \varepsilon I$ and every $r > 0$.*

Let $a \varepsilon C(0)$ and consider the respective Yosida approximants u_λ $(\lambda > 0)$.

a) *When $\lambda \to 0$, u_λ converges pointwisely to w^- , where w is the solution to the sweeping process (2)-(4) (see Definition 1.1):*

(9) $\lim\limits_{\lambda \to 0} \ \| u_\lambda(t) - w^-(t) \| = 0 \quad (t \varepsilon I)$;

b) *When $\lambda \to 0$, u_λ converges to w in the sense of filled-in graphs, that is :*

(10) $h(gr \ u_\lambda \ , \ gr^* \ w) \to 0$;

c) *When $\lambda \to 0$, there is a strong convergence, uniformly on I:*

(11) $\mathrm{proj}\,(u_\lambda(t) \, , \ C(t)) \to w(t) \quad (t \varepsilon I)$.

     The proof runs as follows. Since $\mathbb{R}$ is metrizable, it suffices to assume that λ takes a sequence of positive values converging to zero. In section 2, we give some estimates on the L^1-norm of the derivatives of the Yosida approximants. This allows extraction of a subsequence which is weakly pointwisely convergent to a bv function u, which turns out to be left-continuous. In section 3, several properties of u and u^+ are established; in particular, *strong* pointwise convergence is derived from hypothesis (8). In section 4, it is shown that u^+ is the solution to the sweeping process and by uniqueness the first part follows. In section 5, we study the graph convergence of (u_λ) and prove (10) and (11) with a similar technique. Notice that theorem 0.3.4 on the extraction of convergent subsequences, in the sense of filled-in graphs, does not apply here, thus forcing some level of complexity on the proof.

1.2. Preliminary estimates

We aim at obtaining an upper bound of the Hilbert norm of $\dfrac{du_\lambda}{dt}\,(t)$ using to that effect some approximations of u_λ .

     Let $v : I \to \mathbb{R}$ be the variation function of C (or more generally a "super-variation", i.e., a function that satisfies (1.1)). By assumption, v has bounded variation (since v is nondecreasing, this is equivalent to $v(T) \varepsilon \mathbb{R}$), it is right-continuous on I and continuous at T: $v^-(T) = v(T)$. Hence, for every $\epsilon > 0$, it is possible to find a partition P_ϵ of the interval I:

$$t_0 = 0 < t_1 < \ldots < t_m < T = t_{m+1}$$

such that, putting $I_i = [t_i, t_{i+1}[$ for $0 \le i < m$ and $I_m = [t_m, T]$, we have, for every i:

(1) $$\text{length}(I_i) \le \epsilon \ , \ \text{osc}(v, I_i) \le \epsilon.$$

To P_ϵ we associate a step-multifunction C_ϵ defined by

$$C_\epsilon(t) = C(t_i) \ , \ \text{if } t \epsilon I_i \ ,$$

which satisfies, for every t:

(2) $$h(C_\epsilon(t), C(t)) = h(C(t_i), C(t)) \le v(t) - v(t_i) \le \epsilon \ .$$

For technical reasons, we shall temporarily consider an arbitrary initial value $a \epsilon H$ and denote by u_λ, respectively by $u_{\epsilon,\lambda}$, the absolutely continuous solution of (1.6)-(1.7), respectively of

(3) $$\frac{du_{\epsilon,\lambda}}{dt}(t) + \frac{1}{\lambda}[u_{\epsilon,\lambda}(t) - \text{proj}(u_{\epsilon,\lambda}(t), C_\epsilon(t))] = 0 \ , \ \text{a. e. on } I$$

with the initial condition $u_{\epsilon,\lambda}(0) = a$.

From the elementary theory of ordinary differential equations follow the equivalent integral formulations [Bre 1]:

(4) $$u_\lambda(t) = e^{-t/\lambda} a + \frac{1}{\lambda} \int_0^t e^{(s-t)/\lambda} \ \text{proj}(u_\lambda(s), C(s)) \, ds \ ;$$

(5) $$u_{\epsilon,\lambda}(t) = e^{-t/\lambda} a + \frac{1}{\lambda} \int_0^t e^{(s-t)/\lambda} \ \text{proj}(u_{\epsilon,\lambda}(s), C_\epsilon(s)) \, ds.$$

The latter can be computed explicitly. Let

$$b = \text{proj}(a, \ C(0)).$$

Since $C_\epsilon(t) = C(0)$ for $0 \le t < t_1$, we guess that, for these values of t, $u_{\epsilon,\lambda}(t)$ must belong to the segment $[a, b]$, i.e., $u_{\epsilon,\lambda}(t) = b + \psi(t)(a-b)$, with $0 \le \psi(t) \le 1$ and $\psi(0) = 1$. This reduces equation (3) to

$$\frac{d\psi}{dt}(t)(a-b) + \frac{1}{\lambda}\psi(t)(a-b) = 0 \quad \text{a. e. ,}$$

since $\text{proj}(u_{\epsilon,\lambda}(t), C(0)) = b$. Solving this Cauchy problem for ψ, in the nontrivial case $a \ne b$, gives $\psi(t) = e^{-t/\lambda}$; so, for $t \epsilon I_0$,

$$u_{\epsilon,\lambda}(t) = b + e^{-t/\lambda}(a-b) \ .$$

Proceeding in the same manner when we consider the other subintervals, it is easily checked that, if $x_i = u_{\epsilon,\lambda}(t_i) = (u_{\epsilon,\lambda})^-(t_i)$ and $y_i = \mathrm{proj}(x_i\,,\,C(t_i))$ (in particular, $x_0 = a$, $y_0 = b$), then, for $t\varepsilon\,I_i$:

$$(6) \qquad u_{\epsilon,\lambda}(t) \;=\; y_i + e^{-(t-t_i)/\lambda}\,(x_i - y_i)\,.$$

We want to show that these functions $u_{\epsilon,\lambda}$ converge to u_λ as $\epsilon \to 0$.

With this aim, we prove the following inequality for $t\varepsilon\,I_i$:

$$(7) \qquad \| u_{\epsilon,\lambda}(t) - \mathrm{proj}(u_{\epsilon,\lambda}(t),C_\epsilon(t)) \| \;\leq\; e^{-t/\lambda}\,\| a - b \|$$
$$+ \sum_{j=1}^{i} e^{-(t-t_j)/\lambda}\,(r(t_j) - r(t_{j-1}))\,,$$

where $r(.)$ is a retraction function of C, or a "super-retraction", meaning that:

$$e(\,C(s)\,,\,C(t)\,) \;\leq\; r(t) - r(s) \quad (s \leq t)\,.$$

Proof of (7). Since $u_{\epsilon,\lambda}$ is continuous, (6) implies

$$\| x_{j+1} - y_j \| \;=\; e^{-(t_{j+1}-t_j)/\lambda}\,\| x_j - y_j \|\,.$$

Moreover, if $j \geq 1$, then since $y_{j-1}\varepsilon\,C(t_{j-1})$ we have

$$\mathrm{dist}(y_{j-1}\,,\,C(t_j)) \;\leq\; e(\,C(t_{j-1}),C(t_j)) \leq r(t_j) - r(t_{j-1}),$$

and so

$$\| x_j - y_j \| \;=\; \mathrm{dist}(x_j,C(t_j)) \;\leq\; \| x_j - y_{j-1} \| + \mathrm{dist}(y_{j-1}\,,\,C(t_j))$$
$$\leq\; e^{-(t_j-t_{j-1})/\lambda}\,\| x_{j-1} - y_{j-1} \| + r(t_j) - r(t_{j-1})\,.$$

Therefore, for $j=1$, $\| x_1 - y_1 \| \leq e^{-t_1/\lambda}\,\| a-b \| + r(t_1) - r(t_0)$; for $j=2$,

$$\| x_2 - y_2 \| \;\leq\; e^{-(t_2-t_1)/\lambda}[\,e^{-t_1/\lambda}\,\| a-b \| + r(t_1) - r(t_0)] + r(t_2) - r(t_1)$$
$$=\; e^{-t_2/\lambda}\,\| a-b \| + e^{-(t_2-t_1)/\lambda}[\,r(t_1) - r(t_0)] + e^{-(t_2-t_2)/\lambda}[r(t_2) - r(t_1)]\,,$$

and by induction

$$\| x_i - y_i \| \;\leq\; e^{-t_i/\lambda}\,\| a-b \| + \sum_{j=1}^{i} e^{-(t_i-t_j)/\lambda}\,[\,r(t_j) - r(t_{j-1})]\,.$$

This inequality implies (7) because, if we consider $t\varepsilon\,I_i$, then by (6) we have:

$$\| u_{\epsilon,\lambda}(t) - \mathrm{proj}(u_{\epsilon,\lambda}(t),C_\epsilon(t)) \| \;=\; \| u_{\epsilon,\lambda}(t) - y_i \| \;=\; e^{-(t-t_i)/\lambda}\,\| x_i - y_i \|\,. \qquad \square$$

Note that in (7):

$$\sum_{j=1}^{i} e^{-(t-t_j)/\lambda}[r(t_j)-r(t_{j-1})] \le \sum_{j=1}^{i} [r(t_j)-r(t_{j-1})] = r(t_i)-r(t_0) \le r(t)-r(0).$$

By definition of the retraction $\mathrm{ret}(C; 0, t)$ as the supremum of sums $\sum e(C(s_i), C(s_{i+1}))$ with $s_0=0 < \ldots < s_n=t$, we have $\mathrm{ret}(C; 0, t) \le r(t)-r(0)$ (with equality if r is really a retraction function). Thus, (7) implies, for every $t \varepsilon I$, the following important estimate:

$$(8) \qquad \| u_{\epsilon, \lambda}(t) - \mathrm{proj}(u_{\epsilon, \lambda}(t), C_\epsilon(t)) \| \le e^{-t/\lambda} \| a-b \| + \mathrm{ret}(C; 0, t)$$

$$\le \| a-b \| + \mathrm{ret}(C; I) := M_1 .$$

For every positive λ, it is also possible to ensure the existence of a constant M_2 – which initially is allowed to vary with λ – such that for every t in I:

$$(9) \qquad \| u_\lambda(t) - \mathrm{proj}(u_\lambda(t), C(t)) \| \le M_2 .$$

In fact, as u_λ is a bounded function on I and recalling the definition of projection and that b belongs to $C(0)$:

$$\| u_\lambda(t) - \mathrm{proj}(u_\lambda(t), C(t)) \| = \mathrm{dist}(u_\lambda(t), C(t)) \le \| u_\lambda(t) - \mathrm{proj}(b, C(t)) \|$$

$$\le \| u_\lambda(t) - b \| + \mathrm{dist}(b, C(t)) \le \| u_\lambda(t) \| + \| b \| + e(C(0), C(t))$$

$$\le \| u_\lambda \|_\infty + \| b \| + \mathrm{ret}(C; I) .$$

We can now prove the following:

Lemma 2.1. *For any $\lambda > 0$, there is a positive constant $L = L(\lambda)$ such that:*

$$(10) \qquad \| u_{\epsilon, \lambda} - u_\lambda \|_\infty \le L(1 + \frac{T}{\lambda}) \sqrt{\epsilon} .$$

Thus, as ϵ tends to zero and λ is fixed, uniform convergence holds:

$$(11) \qquad \| u_{\epsilon, \lambda} - u_\lambda \|_\infty \to 0 , \quad \text{when } \epsilon \to 0 .$$

Proof. A geometrical inequality on projections (Prop. 0.4.7) gives here, upon use of (2), (8) and (9):

$$(12) \quad \| \mathrm{proj}(u_{\epsilon, \lambda}(s), C_\epsilon(s)) - \mathrm{proj}(u_\lambda(s), C(s)) \|^2 \le \| u_{\epsilon, \lambda}(s) - u_\lambda(s) \|^2$$

$$+ 2 h(C_\epsilon(s), C(s)) [\mathrm{dist}(u_{\epsilon, \lambda}(s), C_\epsilon(s)) + \mathrm{dist}(u_\lambda(s), C(s))]$$

$$\le \| u_{\epsilon, \lambda}(s) - u_\lambda(s) \|^2 + 2 (M_1 + M_2) \epsilon .$$

Let $\phi(t) := \| u_{\epsilon, \lambda}(t) - u_\lambda(t) \|$, $L := \sqrt{2(M_1 + M_2)}$ and $K := L \sqrt{\epsilon}$. By (4), (5) and (12) we are led to:

$$\phi(t) \leq \frac{1}{\lambda} \int_0^t e^{(s-t)/\lambda} \sqrt{\phi(s)^2 + L^2\epsilon}\; ds,$$

hence,

$$\phi(t) \leq L\sqrt{\epsilon}\,(1 - e^{-t/\lambda}) + \int_0^t \frac{1}{\lambda}\, e^{(s-t)/\lambda}\phi(s)\,ds \leq K + \psi(t),$$

where $\psi(t)$ denotes the last integral. Since $\psi(0) = 0$ and

$$\frac{d\psi}{dt} = \frac{1}{\lambda}\,(\phi - \psi) \leq \frac{K}{\lambda} \quad \text{a. e. on } I,$$

it follows that

$$\phi(t) \leq K + \frac{K}{\lambda}\,t = L\left(1 + \frac{t}{\lambda}\right)\sqrt{\epsilon}\,,$$

whence (10). $\qquad\qquad\qquad\qquad\qquad\qquad\qquad\qquad\qquad\qquad\qquad\qquad\quad\square$

Denoting by dr the differential or Stieltjes measure of the retraction function r, we state:

Theorem 2.2. *For every t in I and every positive λ, the following estimate holds:*

$$(13) \quad \| u_\lambda(t) - \mathrm{proj}(u_\lambda(t), C(t)) \| \leq e^{-t/\lambda}\,\| a - b \| + \int_{[0,\,t]} e^{(s-t)/\lambda}\, dr(s)$$

$$\leq \| a - b \| + \mathrm{ret}(C; 0, t).$$

Proof. When $\epsilon \to 0$, $C_\epsilon(t)$ converges to $C(t)$ in the sense of Hausdorff distance, as is clear from (2), whereas by (11), $u_{\epsilon,\lambda}(t)$ converges strongly in H to $u_\lambda(t)$. Hence the left-hand side of inequality (7) converges to the left-hand side of (13) (see (12) or use the continuity of the projection into a fixed convex set). We further remark that the sum appearing in the right-hand side of (7) is bounded above by

$$\sum_{j=1}^{i} e^{(t_j - t)/\lambda}[r(t_j) - r(t_{j-1})] + e^{(t-t)/\lambda}\,[r(t) - r(t_j)]\,.$$

This is a Stieltjes sum for the integral over $[0, t]$ of the continuous function $s \to e^{(s-t)/\lambda}$ with respect to the differential measure dr. Since by construction the subdivision $P_\epsilon = \{ t_0, \ldots, t_{m+1} \}$ satisfies (1), we have $dr(\mathrm{int}\, I_i) = \mathrm{osc}(r, I_i) \leq \mathrm{osc}(v, I_i) \leq \epsilon$ and these Stieltjes sums converge to the said integral when ϵ tends to zero. So, inequality (7) implies the first inequality in (13).

The second inequality in (13) is now immediate, because

$$\int_{[0,\,t]} e^{(s-t)/\lambda}\, dr(s) \leq \int_{[0,\,t]} dr(s) = r(t) - r(0). \qquad\qquad\quad\square$$

The following estimate is fundamental:

Theorem 2.3. *The Yosida approximants satisfy, for every positive λ, the inequality:*

$$(14) \qquad \int_0^T \| \frac{du_\lambda}{dt} \| \, dt \leq \| a - b \| + \mathrm{ret}(C; 0, T) \, .$$

Proof. Thanks to (1.6) and (13):

$$\int_0^T \| \frac{du_\lambda}{dt} \| \, dt \ \leq \ \frac{1}{\lambda} \int_0^T [\, e^{-t/\lambda} \| a - b \| + \int_{[0,\,t]} e^{(s-t)/\lambda} \, dr(s)] \, dt$$

$$= (1 - e^{-T/\lambda}) \| a - b \| + \frac{1}{\lambda} \int_{[0,\,T]} dr(s) \int_s^T e^{(s-t)/\lambda} \, dt$$

$$\leq \| a - b \| + \int_{[0,\,T]} [1 - e^{(s-T)/\lambda}] \, dr(s)$$

$$\leq \| a - b \| + \int_{[0,\,T]} dr(s) \, . \qquad\qquad \square$$

We may replace I by a subinterval $[s,\ t]$ with $s \leq t$. The restriction of u_λ to the latter is then the unique solution to (1.6) taking the value $u_\lambda(s)$ at the initial instant s. Applying the above results, it turns out that we have proved more generally that

$$(15) \qquad \| u_\lambda(t) - u_\lambda(s) \| \ \leq \ \int_s^t \| \frac{du_\lambda}{d\tau} \| \, d\tau$$

$$\leq \ \| u_\lambda(s) - \mathrm{proj}(u_\lambda(s),\ C(s)) \| + \mathrm{ret}(C;\ s,\ t) \, .$$

1.3. Limit functions

From now on it is assumed that $a \in C(0)$ and so $b = \mathrm{proj}(a, C(0)) = a$. Then inequality (2.14) gives:

$$(1) \qquad \int_0^T \| \frac{du_\lambda}{dt} \| \, dt \ \leq \ \mathrm{ret}(C; 0, T)$$

and since $u_\lambda(t) = a$ it follows that for every λ and every t:

$$(2) \qquad \| u_\lambda(t) \| \ \leq \ \| a \| + \mathrm{ret}(C; 0, T) \, .$$

Let λ assume any sequence of positive values converging to zero, e. g. $\lambda = 1/n$. The corresponding Yosida approximants form a sequence (u_λ) of uniformly bounded functions with uniformly bounded variation, thanks to (2) and (1) respectively, and which take their values in the Hilbert space H. Then Theorem 0.2.1 ensures the existence of a subsequence $\mathcal{G}$ (in the usual strict sense) of the initial λ sequence and of a function of bounded variation $u\colon I \to H$ such that, for every t, in the weak topology of H:

$$(3) \qquad \text{w-}\lim_{\mathcal{G}} u_\lambda(t) \;=\; u(t).$$

We now proceed to study this limit function u and its right-continuous regularization u^+, which has also bounded variation. The first result is given in a more general setting than needed in the sequel:

Lemma 3.1. *Let* r, *the retraction function of* C, *be continuous at some point* t *in* $[0, T]$. *Then:*

$$(4) \qquad \lim_{\lambda \to 0} \; \| \, u_\lambda(t) - \mathrm{proj}(u_\lambda(t), C(t)) \, \| \; = \; 0 \, ,$$

$$(5) \qquad u(t) \, \varepsilon \, C(t) \, ,$$

$$(6) \qquad \lim_{\mathcal{G}} \; \| \, u_\lambda(t) - u(t) \, \| \; = \; 0 \, .$$

Proof. If $t = 0$, then everything is obvious, because $u_\lambda(0) \; = \; u(0) \; = \; a \, \varepsilon \, C(0)$.

Let $t > 0$. Since $a = b$ and by hypothesis t is not an atom of the measure dr, we may write the inequality (2.13) as

$$\| \, u_\lambda(t) - \mathrm{proj}(u_\lambda(t), C(t)) \, \| \; \leq \; \int_{[0,\, t[} e^{(s-t)/\lambda} \, dr(s) \, ;$$

these integrands are uniformly bounded and for $s \varepsilon [0, t[$ they converge pointwisely to zero as $\lambda \to 0$. By Lebesgue's theorem on dominated convergence, we get (4).

Comparing (3) and (4), it turns out that $u(t)$ is also the weak limit of the sequence $v_\lambda(t) = \mathrm{proj}(u_\lambda(t), C(t))$, with $\lambda \varepsilon \mathcal{G}$. Since this sequence is contained in the weakly closed convex set $C(t)$, (5) follows.

On the other hand, the weakly convergent sequence $v_\lambda(t)$ is bounded in $C(t)$ and so is relatively strongly compact, by virtue of assumption (1.8). Hence, it converges strongly to $u(t)$. Together with (4), this implies (6). $\qquad \square$

The points where the retraction r is continuous form a dense subset of I. Thus, every t in $[0, T[$ may be approximated from the right by a sequence (t_n) of such points, for which we have shown that $u(t_n) \varepsilon C(t_n)$. Since C is a right-continuous multifunction, we deduce that:

$$(7) \qquad u^+(t) \varepsilon C(t) \qquad (t \varepsilon [0,\, T]),$$

where by convention $u^+(T) := u(T)$.

Another relevant consequence is given in the following proposition (which also shows that u^+ has bounded variation):

Proposition 3.2. *For every* $0 \le s \le t \le T$:

$$(8) \qquad \| u^+(t) - u^+(s) \| \le \mathrm{ret}(C; s, t) = r(t) - r(s).$$

Proof. In the nontrivial case $s < t$, assume that t' and t'' are two continuity points of r such that $s \le t' < t \le t''$. Using (2.15) leads to

$$\| u_\lambda(t'') - u_\lambda(t') \| \le \| u_\lambda(t') - \mathrm{proj}(u_\lambda(t'), C(t')) \| + r(t'') - r(t').$$

In view of (4) and (6), this implies

$$\| u(t'') - u(t') \| \le r(t'') - r(t') .$$

Letting $t' \to s$ and $t'' \to t$ and recalling that r is right-continuous, we obtain (8). $\qquad\qquad\qquad\qquad\qquad\qquad\qquad\qquad\qquad\qquad\qquad\qquad\quad\square$

Notice that if r is continuous at s, then in the preceding proof we may take $t' = s$ and $t'' \to s$, thus establishing that

$$(9) \qquad u^+(s) = u(s), \text{ whenever } r \text{ is continuous at } s.$$

In particular,

$$(10) \qquad u^+(0) = u(0) = a .$$

Since u^+ and r are both rcbv functions, $u^+(t) - u^+(s) = \int_{]s,t]} du^+$ and $r(t) - r(s) = \int_{]s,t]} dr$. Then (8) implies the following inequality about the measure of total variation of du^+:

$$(11) \qquad | du^+ | \le dr ,$$

in the sense of the ordering of (positive) real measures.

Proposition 3.3. *The pointwise limit function u is left-continuous:*

$$(12) \qquad\qquad u^-(t) = u(t) \qquad (t \in I)$$

and strong convergence holds:

$$(13) \qquad\qquad \lim_{\mathscr{I}} \| u_\lambda(t) - u(t) \| = 0 \qquad (t \in I).$$

Proof. In view of (3) and by uniqueness of limits we only need to prove that:

$$(14) \qquad\qquad \lim_{\mathscr{I}} \| u_\lambda(t) - u^-(t) \| = 0 \qquad (t \in I).$$

If $t = 0$, (12) holds by convention and (14) is obvious. If $t \in \,]0, T]$, given an arbitrary $\epsilon > 0$ we choose $s < t$ such that r is continuous at s and $r^-(t) - r(s) \le \epsilon$. By Lemma 3.1, for sufficiently small λ in the subsequence $\mathscr{I}$ the following inequalities hold simultaneously:

$$\| u_\lambda(s) - u(s) \| \le \epsilon \,, \quad \| u_\lambda(s) - \mathrm{proj}(u_\lambda(s), C(s)) \| \le \epsilon \,.$$

Then, for any t' in $[s, t[\,$, recalling (2.15):

$$\| u_\lambda(t') - u_\lambda(s) \| \le \epsilon + r(t') - r(s) \le \epsilon + r^-(t) - r(s) \le 2\epsilon.$$

Together with (8) and (9) this now gives:

$$(15) \quad \| u_\lambda(t') - u^+(t') \| \le \| u_\lambda(t') - u_\lambda(s) \| + \| u_\lambda(s) - u(s) \| + \| u^+(s) - u^+(t') \|$$

$$\le 3\epsilon + r(t') - r(s) \le 4\epsilon \,.$$

Fixing λ in the afore-mentioned conditions, we take limits as $t' \to t$, obtaining:

$$\| u_\lambda(t) - u^-(t) \| \le 4\epsilon \,,$$

because u_λ is continuous and $(u^+)^- = u^-$. Hence (14) holds. $\qquad\square$

From (8) and (12) we readily infer that:

$$(16) \qquad\qquad \| u^+(t) - u(t) \| \le r(t) - r^-(t).$$

We end this section with a statement which is more precise than (7):

Proposition 3.4. *For every t in I, we have*

$$(17) \qquad\qquad u^+(t) = \mathrm{proj}(u(t), C(t)) \,.$$

Proof. If $t = T$, this is clear by (7) and the convention thereafter. If $t < T$, let $t' > t$. By (2.15), $\| u_\lambda(t') - u_\lambda(t) \| \le \mathrm{dist}(u_\lambda(t), C(t)) + \mathrm{ret}(C; t, t')$.
Taking limits with respect to the subsequence $\mathscr{I}$ and profiting from (13) and

from the strong continuity of the distance to a fixed set, we get:

$$\| u(t') - u(t) \| \leq \mathrm{dist}(u(t), C(t)) + r(t') - r(t) .$$

Letting t' converge to t and remembering that r is right-continuous, this implies that

$$\| u^+(t) - u(t) \| \leq \mathrm{dist}(u(t), C(t)) .$$

Since $u^+(t) \varepsilon C(t)$ (by (7)), this characterizes $u^+(t)$ as the projection of $u(t)$. $\square$

1.4. The solution

We shall prove that $w := u^+$ *is the solution to the sweeping process* (1.2)-(1.3) *with initial value* a.

We know that w has bounded variation (3.8) and is of course right-continuous. In view of (3.10), it satisfies the prescribed initial condition and (3.7) — or (3.17) — means that $w(t) \varepsilon C(t)$ for every t in I. Hence all we have to show is (1.4), that is, choosing $d\mu = | du^+ |$:

$$(1) \qquad\qquad - \frac{du^+}{| du^+ |}(t) \varepsilon N_{C(t)}(u^+(t)) ,$$

for $| du^+ |$-almost every t in I. Notice that the measures du and du^+ are equal on I and have no atoms at the endpoints $t = 0$ and $t = T$, because of (3.10) and since $u^-(T) = u(T)$, by (3.12).

If $t \varepsilon \,]0, T[$ is an atom of $du^+ = du$, we have by (3.12) and (3.17):

$$- \frac{du^+}{| du^+ |}(t) \;=\; - \frac{u^+(t) - u^-(t)}{\| u^+(t) - u^-(t) \|} \;=\; \frac{u(t) - \mathrm{proj}(u(t), C(t))}{\| u(t) - \mathrm{proj}(u(t), C(t)) \|} .$$

By the property of projections, this vector belongs to the outward normal cone to the convex set $C(t)$ at the point $u^+(t) = \mathrm{proj}(u(t), C(t))$. So, we have established that (1) holds at every atom.

Next, we shall consider the points at which u^+ (and hence u) is continuous. To be precise, because (1) is to be verified only du^+-almost everywhere, we only need to consider the points t where the retraction r is continuous. This excludes a countable set, possibly formed by: 1) the atoms of $du = du^+$, already studied, and 2) a subset of continuity points for u^+ (hence du^+-null points) which is then du^+-null itself.

The following lemma is used to obtain (1).

Lemma 4.1. *If an rcbv function $\phi : [t, t'] \to H$ is a selection of C (i.e., $\phi(s) \, \varepsilon \, C(s)$ for every s in the interval) then:*

$$(2) \qquad\qquad \int_{[t,\,t'[} \phi \cdot du \geq \tfrac{1}{2} \left(\| u(t') \|^2 - \| u(t) \|^2 \right) .$$

Proof. Recall that here $\int_{[t,t'[} \phi \cdot du := \int_{[t,t'[} \left(\phi \cdot \dfrac{du}{|du|} \right) |du| \, \varepsilon \, \mathbb{R}$, where the density $\dfrac{du}{|du|}$ belongs to $\mathcal{L}^1(I, |du| ; H)$.

We still use the notation $v_\lambda(s) = \operatorname{proj}(u_\lambda(s), \, C(s))$. By definition (1.6), the Yosida approximants satisfy, for (Lebesgue) almost every s:

$$\frac{du_\lambda}{ds}(s) \;=\; \tfrac{1}{\lambda} \left(v_\lambda(s) - u_\lambda(s) \right) .$$

Since $\phi(s) \, \varepsilon \, C(s)$ and λ is positive, this implies by a well-known property of projections:

$$[\phi(s) - v_\lambda(s)] \cdot \frac{du_\lambda}{ds}(s) \geq 0 , \qquad \text{a.e. in } [t, t'].$$

Then

$$\int_t^{t'} \phi \cdot du_\lambda = \int_t^{t'} \phi \cdot \frac{du_\lambda}{ds} \, ds \geq \int_t^{t'} v_\lambda \cdot \frac{du_\lambda}{ds} \, ds$$

and since

$$v_\lambda \cdot \frac{du_\lambda}{ds} = (v_\lambda - u_\lambda) \cdot \frac{du_\lambda}{ds} + u_\lambda \cdot \frac{du_\lambda}{ds} = (v_\lambda - u_\lambda) \cdot \left[\tfrac{1}{\lambda}(v_\lambda - u_\lambda) \right] + u_\lambda \cdot \frac{du_\lambda}{ds}$$

it follows that

$$v_\lambda \cdot \frac{du_\lambda}{ds} \geq u_\lambda \cdot \frac{du_\lambda}{ds}$$

and

$$\int_t^{t'} \phi \cdot du_\lambda \geq \int_t^{t'} u_\lambda \cdot \frac{du_\lambda}{ds} \, ds = \tfrac{1}{2} \left(\| u_\lambda(t') \|^2 - \| u_\lambda(t) \|^2 \right) .$$

We take limits with respect to λ. Since the continuous functions u_λ converge pointwisely weakly to the left-continuous function u and since ϕ is right-continuous with bounded variation, we may apply Theorem 0.2.1.(ii) to the left-hand side of the last inequality, obtaining:

$$\int_t^{t'} \phi \cdot du_\lambda = \int_{[t,\,t'[} \phi \cdot du_\lambda \;\to\; \int_{[t,\,t'[} \phi \cdot du .$$

Recalling (3.13) yields the desired property. $\qquad\qquad\qquad\qquad \square$

Let t be a continuity point of the retraction r.

By (3.8), u^+ is also continuous at t. For every x in the convex set $C(t)$ we pick an rcbv selection of C defined in $[t,T]$, say ϕ, such that $\phi(t) = x$. This can be done, for instance, by considering the sweeping process with this initial condition and applying the results obtained so far. Notice that the "obvious" choice $s \to \mathrm{proj}(x, C(s))$ may have unbounded variation. By (2), if $\epsilon > 0$ is sufficiently small, we have

$$\int_{[t,\,t+\epsilon[} \phi \cdot du \geq \tfrac{1}{2}(\,\| u(t+\epsilon)\,\|^2 - \| u(t) \|^2\,) = \tfrac{1}{2}[u(t+\epsilon) + u(t)] \cdot [u(t+\epsilon) - u(t)].$$

We write $\phi = x - (x - \phi)$ and choose ϵ in such a way that u is also continuous at $t+\epsilon$, so that $du([t,t+\epsilon[) = u(t+\epsilon) - u(t) = du([t,t+\epsilon])$. We get

$$x \cdot du([t,t+\epsilon]) \geq \tfrac{1}{2}[u(t+\epsilon) + u(t)] \cdot du([t,t+\epsilon]) + \int_{[t,\,t+\epsilon]} (x - \phi) \cdot du,$$

whence

$$(x - \tfrac{1}{2}[u(t+\epsilon) + u(t)]\,) \cdot du([t,t+\epsilon]) \geq -\,\mathrm{osc}(\phi,[t,\,t+\epsilon])\,|\,du\,|\,([t,\,t+\epsilon])\,.$$

Divide by $|\,du\,|\,([t,\,t+\epsilon])$ (with the convention $0/0 = 0$) and let $\epsilon \to 0$, picking only continuity points $t+\epsilon$. By virtue of Jeffery's theorem on the densities of measures (Theorem 0.1.1), this gives:

$$\lim_{\epsilon} (x - \tfrac{1}{2}[u(t+\epsilon) + u(t)]) \cdot \frac{du}{|\,du\,|}(t) \geq -\lim_{\epsilon} \mathrm{osc}(\phi,[t,\,t+\epsilon]),$$

for $|\,du\,|$-almost every t satisfying the above conditions. So, because u is continuous at t and ϕ is right-continuous:

$$(x - u(t)) \cdot \frac{du}{|\,du\,|}(t) \geq 0\,,$$

for every x in $C(t)$. This is equivalent to

$$-\frac{du}{|\,du\,|}(t) \varepsilon\, N_{C(t)}(u(t))\,,$$

which expresses precisely the same as (1), since t is a continuity point.

In short, we have shown so far that, if λ takes *any* sequence of positive values converging to zero and if $\mathcal{I}$ is *any* subsequence ensuring the pointwise weak convergence of the Yosida approximants (u_λ) with $\lambda \varepsilon \mathcal{I}$, then these approximants converge pointwisely *strongly* to $u = w^-$ (see (3.13)) where $w = u^+$ is the *unique* solution to the sweeping process. A standard reasoning shows that, for every t and every $x \varepsilon H$, $w^-(t) \cdot x$ is the unique sublimit and hence the limit of the relatively compact sequence of real numbers $(u_\lambda(t) \cdot x)$.

In other words, the initial sequence (u_λ) converges pointwisely weakly to u and, as mentioned above, it converges also pointwisely strongly. Since the sequence itself was arbitrarily chosen, this amounts to having established (1.9):

$$\lim_{\lambda \to 0} \; \| u_\lambda(t) - w^-(t) \| = 0 \quad (t \varepsilon I) .$$

1.5. Graph convergence

We now show that (u_λ) converges in the sense of filled-in graphs to $w = u^+$, the solution to the sweeping process. Since the Yosida approximants are continuous, their filled-in graphs gr^*u_λ coincide with their graphs given by $gr\, u_\lambda = \{(t,\, u_\lambda(t)) : t \varepsilon I\}$. Since u is left-continuous (Proposition 3.3):

$$gr^*w \; = \; gr^*u^+ \; = \; \{ \, (t, x) : \; t \varepsilon I \text{ and } x \varepsilon [u(t), u^+(t)] \, \} .$$

Theorem 5.1. *When $\lambda \to 0$, the Yosida approximants u_λ converge to $w = u^+$ in the sense of filled-in graphs, that is:*

$$h^*(u_\lambda, w) \; = \; h(gr\, u_\lambda, \, gr^*u^+) \to 0 .$$

Proof. Let $M' \geq \| u^+ - u \|_\infty$; for instance, take $M' = \| dr \|$, by (3.16). It suffices to prove that, given any ϵ in $]0, 1/6[$, the following estimate holds for sufficiently small positive λ :

$$(1) \qquad\qquad h(gr\, u_\lambda, \, gr^*u^+) \leq M(\epsilon) := (3 + M')\sqrt{6\epsilon} .$$

Take a partition I_0 , I_1 , ... , I_m which satisfies (2.1): the lengths of the subintervals and the respective oscillations of v are bounded above by ϵ. Reasoning for a fixed interval I_i, we choose a continuity point for the retraction t'_i in $]t_i, t_{i+1}[$ and a positive number λ_i such that for every $\lambda \varepsilon]0, \lambda_i]$ we have:

$$(2) \qquad \| u_\lambda(t'_i) - u(t'_i) \| \leq \epsilon, \; \| u_\lambda(t_i) - u(t_i) \| \leq \epsilon,$$

$$\| u_\lambda(t'_i) - \mathrm{proj}(u_\lambda(t'_i),\, C(t'_i)) \| \leq \epsilon ;$$

this is made possible by the strong pointwise convergence of the Yosida approximants and by (3.4).

Proceeding as in the proof of Proposition 3.3, inequality (3.15), we obtain for those values of λ :

(3) $$\| u^+(t) - u_\lambda(t) \| \leq 4\epsilon \qquad (t \in [t'_i, t_{i+1}[) \, ,$$

while, for every $t > t_i$ in I_i , (3.16) implies that:

(4) $$\| u^+(t) - u(t) \| \leq r(t) - r^-(t) \leq \epsilon \, .$$

Then, if $0 < \lambda \leq \lambda_i$ and $t'_i \leq t < t_{i+1}$, we have, by (3) and (4):

(5) $$\| x - u_\lambda(t) \| \leq 5\epsilon, \text{ for any } x \in [u(t), u^+(t)] \, .$$

If $t \in [t_i, t'_i]$, we use instead (2.1), (3.8), (3.9) and (2) in order to get

(6) $$\| u^+(t) - u_\lambda(t'_i) \| \leq \| u^+(t) - u^+(t'_i) \| + \| u(t'_i) - u_\lambda(t'_i) \| \leq 2\epsilon \, .$$

Together with (2.1) and (4) this gives, for $x \in [u(t), u^+(t)]$ and $t \in \,]t_i, t'_i]$:

(7) $$\delta((t, x), (t'_i, u_\lambda(t'_i))) := \max\{ \, | t - t'_i | \, , \, \| x - u_\lambda(t'_i) \| \, \} \leq \max\{\epsilon, 3\epsilon\} = 3\epsilon \, .$$

Now consider the absolutely continuous path

$$l := \{ \, u_\lambda(t) \colon \ t \in [t_i, t'_i] \, \} \, .$$

Its endpoints $u_\lambda(t_i)$ and $u_\lambda(t'_i)$ satisfy respectively (2) and (6), where in the latter we take $t = t_i$; in short, they are not very far from the endpoints of the line segment $S = [u(t_i), u^+(t_i)]$. Let us also compare the length of l, $|\,l\,|$, with the length of S. Using (2.15) and then (2.1), Proposition 3.4, (2) and the non-expansiveness of $x \to \mathrm{proj}(x, C(t_i))$, we obtain the estimates:

$$| \, l \, | \leq \ \| u_\lambda(t_i) - \mathrm{proj}(u_\lambda(t_i), C(t_i)) \| + r(t'_i) - r(t_i)$$

$$\leq \| u_\lambda(t_i) - u(t_i) \| + \| u(t_i) - u^+(t_i) \|$$

$$+ \| \mathrm{proj}(u(t_i), C(t_i)) - \mathrm{proj}(u_\lambda(t_i), C(t_i)) \| + \epsilon$$

$$\leq 2 \| u_\lambda(t_i) - u(t_i) \| + \| u(t_i) - u^+(t_i) \| + \epsilon,$$

$$| \, l \, | \leq \| u(t_i) - u^+(t_i) \| + 3\epsilon \, .$$

Then, applying Lemma 0.4.1 to l and S clearly yields:

(8) $$h(l, S) \leq (3 + \| u(t_i) - u^+(t_i) \|) \sqrt{3\epsilon + \epsilon + 2\epsilon} \leq (3 + M') \sqrt{6\epsilon} = M(\epsilon) \, .$$

Finally, we remark that (5), (7) and (8) imply that if $0 < \lambda \leq \lambda_i$, then:

$$h(\, gr \, u_\lambda \cap (I_i \times H) \, , \ gr^* u^+ \cap (I_i \times H)) \ \leq \ \max\{ 5\epsilon, \, 3\epsilon, \, M(\epsilon) \} = M(\epsilon) \, .$$

Thus (1) is satisfied for every positive $\lambda \leq \min \{\lambda_0, \ldots, \lambda_m\}$. $\qquad \square$

The same technique leads also to our last result concerning this problem:

Theorem 5.2. *The projections of the Yosida approximants, denoted by* $v_\lambda(t) := \mathrm{proj}(u_\lambda(t),\, C(t))$, *converge uniformly in the norm of H to the solution to the sweeping process $w = u^+$, as $\lambda \to 0$.*

Proof. Let $0 < \epsilon < \frac{1}{6}$ be given and choose any partition (I_i) of the interval I satisfying (2.1). We have seen that (3) and (8) hold for every i and every sufficiently small λ. We consider two cases separately.

If $t \varepsilon [t'_i, t_{i+1}[$, then by (3.7) and (3):

$$
(9) \qquad \| v_\lambda(t) - u^+(t) \| = \| \mathrm{proj}(u_\lambda(t), C(t)) - \mathrm{proj}(u^+(t), C(t)) \|
$$

$$
\leq \| u_\lambda(t) - u^+(t) \| \leq 4\epsilon .
$$

Let now $t\varepsilon[t_i, t'_i]$. In view of (8), $u_\lambda(t)$ is not far from the line segment S; to be precise, there exists $x \varepsilon [u(t_i), u^+(t_i)]$ such that

$$
\| u_\lambda(t) - x \|^2 \leq 6(3 + M')^2\, \epsilon .
$$

But $u^+(t_i)$, which by (3.17) is the projection of $u(t_i)$ in $C(t_i)$, is clearly also the projection of x. Hence, by application of (2.2), (2.13), (3.8) and of the inequality on projections (Prop. 0.4.7):

$$
\| v_\lambda(t) - u^+(t) \| \leq \| \mathrm{proj}(u_\lambda(t), C(t)) - \mathrm{proj}(x, C(t_i)) \| + \| u^+(t_i) - u^+(t) \|
$$

$$
\leq \sqrt{ \| u_\lambda(t) - x \|^2 + 2\, h(C(t), C(t_i)) \, [\, \| u_\lambda(t) - v_\lambda(t) \| + \| x - u^+(t_i) \| \,] } + \epsilon
$$

$$
\leq \sqrt{ 6(3 + M')^2\, \epsilon + 2\epsilon\, [\, \| dr \| + M' \,] } + \epsilon .
$$

This is an upper bound which is independent of λ and which converges to zero as $\epsilon \to 0$.

From this and (9) the result is readily obtained. $\qquad\qquad \square$

Chapter 2

Sweeping Processes by Convex Sets with Nonempty Interior

2.1. Introduction

In this Chapter, we shall deal with the sweeping process (Definition 1.1.1) by a moving convex set $t \to C(t)$ with nonempty interior. Sometimes the convex set $C(t)$ may be decomposed in the form

$$C(t) = v(t) + \Gamma(t) \, ,$$

where v is a function taking values in a separable Hilbert space H and Γ is a multifunction with closed convex values having nonempty interior in H. If v is continuous and Γ is Lipschitz-continuous in the sense of Hausdorff distance, then the first proof of existence of a solution is due to Castaing ([Cas 1] Th. 6). It generalizes a previous statement by Tanaka [Tan], where Γ is constant and H is finite-dimensional. Since C need not have a bounded retraction, the assumption on the interior of the convex set is essential. For instance, if we take $\Gamma(t) \equiv \{0\}$, then only the function v could be a solution to the sweeping process, but v may have unbounded variation.

More generally, we establish here the existence of a solution in the following two cases:

a) $C(t)$ is Hausdorff-continuous and H is a Hilbert space of arbitrary dimension (sections 2 and 3);

b) $C(t)$ is lower semicontinuous (from the right) and H is finite-dimensional (section 4).

Uniqueness follows from the monotonicity of the outward normal cone; see [Mor 1, 5] and [Mor 6] (Prop. (6.b)). A random or parametric version of the sweeping process is also studied (Theorem 3.8).

In the implicit kynematical interpretation of this mathematical problem, the fundamental result achieved is that, in both the above cases, *the driven (swept) system has a bounded variation motion even if the driving (sweeping) system has not.*

Proofs are based on the catching-up algorithm conveniently adapted to this new type of assumption.

2.2. Continuous convex set in arbitrary dimension: preliminary results

Consider $I = [0, T]$, a real Hilbert space H with arbitrary dimension and a multifunction $t \to C(t)$ defined on I whose values are closed convex subsets of H with nonempty interior. It is assumed that C is **Hausdorff-continuous**, i. e., continuous in the sense of the Hausdorff distance h relative to the metric associated with the Hilbert norm of H. This means that, for any s in I:

$$(1) \qquad h(\, C(t), C(s)) \to 0 \;, \; \text{as} \;\; t \to s.$$

Note that this condition does not require that the convex sets be bounded, but clearly implies that $h(\, C(t), C(s))$ be finite for every s and t.

We prove that the sweeping process by such a moving convex set C has a solution.

Theorem 2.1. *Under the given hypotheses, for every initial value $u_0 \, \varepsilon \, C(0)$ there is one and only one cbv (continuous with bounded variation) function $u: I \to H$ such that:*

$$(2) \qquad u(0) = u_0 \;;$$

$$(3) \qquad u(t) \, \varepsilon \, C(t) \quad (\, t \, \varepsilon \, I) \;;$$

$$(4) \qquad - \frac{du}{|\, du\,|}(t) \; \varepsilon \; N_{C(t)}\,(u(t)) \;, \quad |\, du\,| \text{- a. e. .}$$

Note that the solution to the sweeping process by $C(t)$ is also the solution to the sweeping process by $C(t) \cap \overline{B}(0, R)$, where $R > \|\, u\,\|_\infty$ (see the proof of Lemma 4.2.2). Hence, if some *a priori* estimate is available, then we may replace the initial multifunction by a bounded one, choosing an appropriate R. This means that assumption (1) may be replaced by the weaker:

$$(5) \qquad t \to s \;\Rightarrow\; h(\, C(t) \cap \overline{B}(0, R), C(s) \cap \overline{B}(0, R)) \to 0 \;,$$

for every positive R.

In the general formulation of sweeping problems, it is only required that the bv solution be right-continuous. The above theorem ensures that the solution is continuous if the moving convex set is Hausdorff-continuous. That this is true can be shown immediately:

Lemma 2.2. *If u is an rcbv function that satisfies (2)-(4) and if $C(.)$ is Hausdorff-continuous, then u is continuous.*

Proof. We assume, by contradiction, that there is some point t in I where u is not continuous, which means that $u^-(t) \neq u(t) = u^+(t)$. Then $t > 0$ is an atom of the measures du and $|du|$ and the density is given at t by:

$$\frac{du}{|du|}(t) = \frac{u^+(t) - u^-(t)}{\|u^+(t) - u^-(t)\|} = \frac{u(t) - u^-(t)}{\|u(t) - u^-(t)\|} \ .$$

Since (4) is satisfied at every atom and its right-hand side is a cone, it follows that $u^-(t) - u(t)$ belongs to the outward normal cone to $C(t)$ at the point $u(t)$. This property characterizes $u(t)$ as the projection of $u^-(t)$ in $C(t)$.

On the other hand, if (t_n) is a nondecreasing sequence converging to t, then we have $u(t_n) \varepsilon \, C(t_n)$, by (3). Thus, we may write:

$$\mathrm{dist}(u(t_n), C(t)) \leq e(C(t_n), C(t)) \leq h(C(t_n), C(t))$$

and pass to the limit, making use of hypothesis (1). This gives $\mathrm{dist}(u^-(t), C(t)) = 0$, i.e., $u^-(t)$ belongs to the closed set $C(t)$. Hence $u(t) = \mathrm{proj}(u^-(t), C(t)) = u^-(t)$, which contradicts the hypothesis and ends the proof. $\qquad\square$

We shall replace (4) by an equivalent condition which is easier to handle. First we remark that the solution to the continuous sweeping process satisfies

$$(6) \qquad\qquad \int_s^t (\phi - u) . \, du \geq 0 \ ,$$

for any $s < t$ in I and for every *continuous selection* of C, $\phi : [s, t] \to H$. (Here there is no need to indicate whether the integration encompasses or not the endpoints of the interval, since du is a nonatomic measure.) Writing $u'(\tau) = \frac{du}{|du|}(\tau)$, we have by hypothesis $-u'(\tau) \varepsilon \, N_{C(\tau)}(u(\tau))$ — except possibly on a $|du|$-null subset — and $\phi(\tau) \varepsilon \, C(\tau)$. By definition of the outward normal cone, we see that

$$-u'(\tau) . [\phi(\tau) - u(\tau)] \leq 0 \ ,$$

hence

$$\int_s^t (\phi - u) . \, du = \int_s^t (\phi(\tau) - u(\tau)) . \, u'(\tau) \, |du|(\tau) \geq 0 \ .$$

It is worth noting that it was precisely this condition (6) that Tanaka and Castaing obtained in their respective problems.

Conversely, we shall prove in Lemma 2.4 that in fact it is sufficient to prove (6) for **constant selections** of C. In the next Lemma, we ensure that enough such selections exist for that purpose.

Lemma 2.3. *Let C be a Hausdorff-continuous multifunction with convex closed values having nonempty interior in a Hilbert space H.*

(a) If $t \varepsilon [0, T[$ and $z \varepsilon \operatorname{int} C(t)$ then there exists $\epsilon > 0$ such that:

$$(7) \qquad z \varepsilon \, C(\tau) \qquad (\, \tau \, \varepsilon \, [t, t+\epsilon] \,) \, .$$

(b) More generally, if $t \varepsilon I$ and $\overline{B}(z, r)$ is a closed ball contained in the interior of $C(t)$, then there exists an interval J_t, open neighbourhood of t in I, such that:

$$(8) \qquad \overline{B}(z, r) \subset C(\tau) \qquad (\tau \, \varepsilon \, J_t) \, .$$

Proof. It is clear that assertion (a) is a consequence of (b): it suffices to take $r = 0$ and to remark that J_t contains some interval $[t, t+\epsilon]$. Let us prove (b). By assumption, the distance d of z to the complement of $C(t)$ is greater than r. Since C is continuous $((1))$, there is an interval J_t which is an open neighbourhood of t in I and such that $h(C(\tau), C(t)) \leq d - r$, for any τ in J_t. Because $C(\tau)$ has nonempty interior, a formula of Moreau (see Prop. 0.4.5) gives in this context:

$$(9) \qquad d = \operatorname{dist}(z, H \backslash \overline{B}(z, d)) \leq \operatorname{dist}(z, H \backslash C(\tau)) + e(\overline{B}(z, d), C(\tau)) \, .$$

But $\overline{B}(z, d) \subset C(t)$, hence

$$e(\overline{B}(z, d), C(\tau)) \, \leq e(C(t), C(\tau)) \leq h(C(t), C(\tau)) \leq d - r \, .$$

So from (9) we deduce that $\operatorname{dist}(z, H \backslash C(\tau)) \geq r$, which implies (8). $\qquad \square$

Lemma 2.4. *Let C be a multifunction with the property (a) of Lemma 2.3 and let $u: I \to H$ be an rcbv function that satisfies (3) and also the following condition:*

$$(10) \qquad z \, . \, [u(t) - u(s)] \geq \tfrac{1}{2} \| u(t) \|^2 - \tfrac{1}{2} \| u(s) \|^2 \, ,$$

if $z \varepsilon \, C(\tau)$ for every $\tau \varepsilon [s, t]$, s and t arbitrary. Then:

$$-\frac{du}{|du|}(t) \ \varepsilon \ N_{C(t)}\ (u(t))$$

holds at $|du|$-almost every point t where u is continuous.

Proof. Jeffery's theorem (Theorem 0.1.1) ensures the existence of a $|du|$-null set $N \subset I$ such that for every $t \varepsilon I \backslash N$ we have simultaneously:

$$(11) \qquad u'(t) = \frac{du}{|du|}(t) = \lim_{\epsilon \to 0+} \frac{du([t,\ t+\epsilon])}{|du|\cdot([t,\ t+\epsilon])}\ ,$$

$$(12) \ \tfrac{1}{2}[u^+(t) + u^-(t)]\cdot u'(t) = \frac{d(\|u\|^2/2)}{|du|}(t) = \lim_{\epsilon \to 0+} \frac{d(\|u\|^2/2)([t,\ t+\epsilon])}{|du|([t,\ t+\epsilon])}\ .$$

Here we use the well-known formula of Moreau for the Stieltjes measure of a quadratic expression of a bv function ([Mor 5] and [Mor 6], (5.7); cf. 0.1(35)):

$$(13) \qquad d(\tfrac{1}{2}\|u\|^2) = \tfrac{1}{2}(u^+ + u^-)\cdot du\ .$$

Now, consider t, a continuity point of u not belonging to $N \cup \{T\}$. By hypothesis (7), every interior point z of $C(t)$ is a constant selection of C in an appropriate interval $[t, t+\epsilon_0]$. Let $(t+\epsilon_n)$ be a sequence of continuity points of u such that $0 < \epsilon_n \le \epsilon_0$ and $\epsilon_n \to 0$. Applying (10) we have:

$$z\cdot[u(t+\epsilon_n) - u(t)] \ge \tfrac{1}{2}\|u(t+\epsilon_n)\|^2 - \tfrac{1}{2}\|u(t)\|^2\ ;$$

equivalently, since we are dealing with endpoints which have zero measure:

$$z\cdot du([t, t+\epsilon_n]) \ge d(\tfrac{1}{2}\|u\|^2)([t, t+\epsilon_n])\ .$$

Dividing by $|du|([t, t+\epsilon_n])$ and taking limits, we obtain

$$z\cdot u'(t) \ge \tfrac{1}{2}[u^+(t) + u^-(t)]\cdot u'(t)\ ,$$

by (11) and (12), whence

$$(z - u(t))\cdot u'(t) \ge 0\ .$$

This inequality, which holds for every z belonging to the interior of the convex set $C(t)$, is easily extended to every z in $C(t)$, by a density argument. Recalling that $u(t) \varepsilon C(t)$, by hypothesis, we find that the inequality expresses precisely that $-u'(t)$ is an outward normal vector to $C(t)$ at $u(t)$. $\qquad \square$

Remark. It is worth noting that any cbv (continuous with bounded variation) function u that satisfies the condition (5) only for constant selections also satisfies assumption (10). In fact, taking z as in Lemma 2.4 and since z is a constant selection of C in the interval $[s, t]$, we have by (5):

$$\int_s^t (z-u)\,.\,du \geq 0 \; ;$$

that is:

$$z\,.\int_s^t du = z\,.\,[u(t)-u(s)] \geq \int_s^t u\,.\,du \;.$$

In the present case, u is assumed continuous and Moreau's formula (13) gives simply:

$$(14) \qquad\qquad d(\tfrac{1}{2}\|u\|^2) = u\,.\,du \;.$$

So:

$$\int_s^t u\,.\,du = \int_s^t d(\tfrac{1}{2}\|u\|^2) = \tfrac{1}{2}\|u(t)\|^2 - \tfrac{1}{2}\|u(s)\|^2$$

and (10) is true. $\qquad\qquad\qquad\qquad\qquad\qquad\qquad\qquad\qquad\qquad\qquad$ $\square$

Combining the preceding results, we may state the following:

Theorem 2.5. *Let C be a Hausdorff-continuous multifunction on the interval I with closed convex values having nonempty interior in the Hilbert space H. Let $u_0 \in C(0)$. Then a bv function $u: I \rightarrow H$ is the solution to the sweeping process by C with initial value u_0 if and only if u is continuous and satisfies (2), (3) and (10).*

2.3. Continuous convex set in arbitrary dimension: algorithm and existence

Let $C: I \rightarrow 2^H$ be a multifunction whose values are closed convex sets with nonempty interior and which is continuous in the sense of Hausdorff distance as in (2.1). Let $u_0 \in C(0)$.

For every $n \geq 1$, we consider the following partition of $I = [0, T]$:

$$(1) \qquad\qquad t_{n,i} = \tfrac{i}{n} T \;\; (0 \leq i \leq n) \;;$$

$$(2) \qquad\qquad I_{n,i} = [t_{n,i}, t_{n,i+1}[\;\; (\text{if } 0 \leq i < n), \;\; I_{n,n} = \{T\}\,.$$

We define $n+1$ elements of H by induction:

$$(3) \qquad\qquad u_{n,0} = u_0 \;;$$

$$(4) \qquad\qquad u_{n,i} = \mathrm{proj}\,(u_{n,i-1}, C(t_{n,i})) \;\; (0 < i \leq n) \;.$$

These $u_{n,i}$ are used to define a right-continuous step-function u_n, which is an approximant of the solution we are looking for:

$$(5) \qquad u_n(t) = u_{n,i} \, , \quad \text{if } t \varepsilon I_{n,i} \quad (0 \leq i \leq n) \, .$$

We also define a continuous piecewise affine approximant v_n, given by:

$$(6) \qquad v_n(t) = u_{n,i} + \frac{t - t_{n,i}}{t_{n,i+1} - t_{n,i}} \, (u_{n,i+1} - u_{n,i}) \quad \text{if } t \varepsilon I_{n,i} \quad (0 \leq i < n)$$

and of course $v_n(T) = u_{n,n}$, by continuity.

We prove the following existence and approximation theorem:

Theorem 3.1. *Both* (u_n) *and* (v_n) *converge uniformly to* u, *the unique solution to the sweeping process* $(2.2) - (2.4)$.

First, we verify that (u_n) and (v_n) do have the same uniform limit, if it exists. In fact, $v_n(T) = u_n(T)$ and for $t \neq T$ it is clear that:

$$\| v_n(t) - u_n(t) \| = \| \frac{t - t_{n,i}}{t_{n,i+1} - t_{n,i}} \, (u_{n,i+1} - u_{n,i}) \| \leq \| u_{n,i+1} - u_{n,i} \|$$

$$= \text{dist}(u_{n,i} \, , \, C(t_{n,i+1})) \leq h(\, C(t_{n,i}), \, C(t_{n,i+1}))$$

and $| t_{n,i} - t_{n,i+1} | = T/n$. Hence

$$(7) \qquad \| v_n - u_n \|_\infty \leq \mu_n \, ,$$

where we put

$$(8) \qquad \mu_n := \sup \{ \, h(C(s), \, C(t)) : \; t, \, s \varepsilon I \, , \; | \, t - s | \leq T/n \, \} \, .$$

Notice that (μ_n) is a nonincreasing sequence that converges to zero:

$$(9) \qquad \lim_n \mu_n \, = \, 0 \, ,$$

thanks to assumption (2.1) and to a classical argument invoking the compactness of the interval I. Combining (9) with (7) we get the expected result.

The next step is to obtain estimates for the functions u_n and their variations (and which also hold for the continuous approximants v_n):

Lemma 3.2. (a) *There exist positive constants* L *and* M *such that, for every* n:

$$(10) \qquad \| u_n(t) \| \leq L \qquad (t \varepsilon I) \, ;$$

$$(11) \qquad\qquad \mathrm{var}\,(u_n\,;\,I) \leq M.$$

(b) *More precisely, if all the convex sets $C(t)$ contain the ball $\overline{B}(a,r)$, we may take*

$$(12) \qquad\qquad L = \|\,u_0\,\| + \|\,u_0 - a\,\|\,,$$

$$(13) \qquad M = l(r, \|\,u_0 - a\,\|\,) := \max\Big\{0, \frac{\|\,u_0 - a\,\|^2 - r^2}{2r}\Big\} \leq \frac{1}{2r}\|\,u_0 - a\,\|^2.$$

Proof. (b) Since $C(t_{n,i}) \supset \overline{B}(a,\,r)$, we deduce from (4) and Lemma 0.4.4 (6) that

$$\|\,u_{n,0} - a\,\| \geq \|\,u_{n,1} - a\,\| \geq \cdots \geq \|\,u_{n,n} - a\,\|\,.$$

Hence $\quad \|\,u_n(t) - a\,\| \leq \|\,u_{n,0} - a\,\| = \|\,u_0 - a\,\|\quad$ and so $\quad \|\,u_n(t)\,\| \leq L,\quad$ with $L = \|\,a\,\| + \|\,u_0 - a\,\|\,.$

On the other hand, from the same Lemma 0.4.4 and 0.4.(8) it follows that

$$\begin{aligned}
\mathrm{var}(u_n\,;\,I) &= \sum_{i=1}^{n} \|\,u_{n,i} - u_{n,i-1}\,\| \\
&\leq \max\Big\{0,\, \frac{1}{2r}\big(\,\|\,u_{n,0} - a\,\|^2 - r^2\,\big)\Big\} = l(r, \|\,u_0 - a\,\|\,) = M.
\end{aligned}$$

(a) By Lemma 2.3.(b), for every t in I we take an interval J_t, which is an open neighbourhood of t in I, such that all the convex sets $C(s)$ with s in J_t contain some fixed closed ball. Since I is a compact set, there is a finite collection of such intervals that still covers I. Then we can find points $s_0 = 0 < s_1 < \cdots < s_p = T$ and closed balls $\overline{B}(a_k, r_k)$ $(k = 1, \ldots, p)$ such that

$$(14) \qquad\qquad C(t) \supset \overline{B}(a_k, r_k) \qquad (t \varepsilon J_k := [s_{k-1}, s_k]\,)\,.$$

From (b), we know that, in J_1, $\|\,u_n(t)\,\| \leq \|\,a_1\,\| + \|\,u_0 - a_1\,\| := L_1$ and $\mathrm{var}(u_n, J_1) \leq \frac{1}{2r_1}\|\,u_0 - a_1\,\|^2 := M_1$.

Hence in J_2 we have

$$\|\,u_n(t)\,\| \leq \|\,u_n(s_1)\,\| + \|\,u_n(s_1) - a_2\,\| \leq 2\,\|\,u_n(s_1)\,\| + \|\,a_2\,\|$$

$$\leq 2\,L_1 + \|\,a_2\,\| := L_2$$

and

$$\mathrm{var}(u_n, J_2) \leq \frac{1}{2r_2}\|\,u_n(s_1) - a_2\,\|^2 \leq \frac{1}{2r_2}(L_1 + \|\,a_2\,\|\,)^2 := M_2\,.$$

In this manner, we obtain constants (not depending on n) $L_1, \ldots, L_p$ and $M_1, \ldots, M_p$. In (10) and (11), we take $L = \max_j\, L_j$ and $M = \max_j\, M_j$. $\qquad\square$

Compactness results for functions of bounded variation such as Theorem 0.2.1 would already enable extraction from (u_n) of a subsequence converging pointwisely weakly to a bv function. However that would not suit our purposes: the much stronger result of uniform convergence of the whole sequence (u_n) is needed. To prove uniform convergence, we shall use the fact that these approximants are actually exact solutions to some sweeping processes by step-multifunctions, which are discretizations of the initial one; with that in mind, we obtain an estimate similar to Moreau's result on the dependence of solutions on the data [Mor 1] (2.16).

Lemma 3.3. *If m is a multiple of n, then for every $t \varepsilon I$ the following inequality holds (with the above notation):*

$$(15) \qquad \| u_n(t) - u_m(t) \|^2 \leq 2\mu_n \left[\operatorname{var}(u_n ; 0, t) + \operatorname{var}(u_m ; 0, t) \right] .$$

Proof. In the first subinterval $I_{m,0} = [0, t_{m,1}[$, (15) is obvious, because $u_n(t) = u_m(t) = u_0$. We suppose by induction that (15) is satisfied in $[0, t_{m,i}[$ and prove it for $t \varepsilon I_{m,i} = [t_{m,i} , t_{m,i+1}[$ $(I_{m,m} = \{T\}$ is dealt with analogously).

Let j be such that $I_{m,i} \subset I_{n,j}$, whence $u_n(t) = u_{n,j}$. There are two possible cases. If $t_{m,i} = t_{n,j}$, then $u_n(t_{m,i-1}) = u_{n,j-1}$ and we have the formula:

$$(16) \qquad u_n(t) = \operatorname{proj}\left(u_n(t_{m,i-1}) , C(t_{n,j}) \right) .$$

If $t_{m,i} \neq t_{n,j}$, then $u_n(t_{m,i-1}) = u_{n,j} \, \varepsilon \, C(t_{n,j})$ and (16) still holds true.

On the other hand, we also have $u_m(t) = u_{m,i} = \operatorname{proj}\left(u_{m,i-1} , C(t_{m,i}) \right)$. Thus, we may apply to $u_n(t)$ and $u_m(t)$ an inequality of Moreau on projections into convex sets (see Proposition 0.4.7):

$$\| u_n(t) - u_m(t) \|^2 \leq \| u_n(t_{m,i-1}) - u_m(t_{m,i-1}) \|^2$$
$$+ 2 \, h(C(t_{n,j}), C(t_{m,i})) \left[\| u_n(t) - u_n(t_{m,i-1}) \| + \| u_m(t) - u_m(t_{m,i-1}) \| \right].$$

By induction hypothesis, (15) holds for $t = t_{m,i-1}$ which belongs to $[0, t_{m,i}[$. Moreover, $h(C(t_{n,j}), C(t_{m,i})) \leq \mu_n$ because $|t_{n,j} - t_{m,i}| \leq T/n$. Then:

$$\| u_n(t) - u_m(t) \|^2 \leq 2\mu_n [\operatorname{var}(u_n ; 0, t_{m,i-1}) + \operatorname{var}(u_m ; 0, t_{m,i-1})]$$
$$+ 2 \mu_n \left[\| u_n(t) - u_n(t_{m,i-1}) \| + \| u_m(t) - u_m(t_{m,i-1}) \| \right]$$

and so

$$\| u_n(t) - u_m(t) \|^2 \leq 2\mu_n \left[\mathrm{var}(u_n;0,t) + \mathrm{var}(u_m;0,t) \right],$$

by definition of variation. $\qquad\qquad\square$

Corollary 3.4. *The approximants (u_n) form a Cauchy sequence in the metric of uniform convergence.*

Proof. In fact, given an arbitrary $\epsilon > 0$, thanks to (9) we pick n such that $\mu_n \leq \epsilon^2/(16\,M)$, where M is an upper bound as in (11). If $p \geq n$, we consider m, the least common multiple of p and m; then the preceding lemma and (11) imply:

$$\| u_n - u_m \|_\infty^2 \leq 4\mu_n M \leq \frac{\epsilon^2}{4}$$

and also

$$\| u_p - u_m \|_\infty^2 \leq 4\mu_p M \leq 4\mu_n M \leq \frac{\epsilon^2}{4} ;$$

whence $\| u_n - u_p \|_\infty \leq \epsilon$, for every $p \geq n$. $\qquad\qquad\square$

So (u_n) converges uniformly to a function $u: I \to H$ and, as we have remarked, the same happens with the continuous approximants (v_n). We shall complete the proof of Theorem 3.1 by establishing the following Lemma.

Lemma 3.5. (a) *The function $u: I \to H$ defined by*:

$$(17) \qquad\qquad u(t) = \lim_n u_n(t) = \lim_n v_n(t)$$

is the cbv solution to the sweeping process (2.2)-(2.4).

(b) *Moreover, if for every t in I, the convex set $C(t)$ contains the ball $\bar{B}(a,r)$, then we have the estimates (see (13))*:

$$(18) \qquad\qquad \mathrm{var}(u;I) \leq l(r, \| u_0 - a \|) \leq \frac{1}{2r} \| u_0 - a \|^2 .$$

Proof. By (11), the uniform limit u must also be a function of bounded variation with $\mathrm{var}(u;I) \leq M$. In particular, (18) follows from (13). As u is the uniform limit of the continuous functions v_n, it is also continuous. Since $u_n(0) = u_{n,0} = u_0$, in the limit we have $u(0) = u_0$, i. e. (2.2).

We write $t \varepsilon D$ if $t = (p/q)\,T$ with $p, q \varepsilon \mathbb{N}$. If n is a multiple of q, then there exists i such that $t = t_{n,i}$ and so

$$u_n(t) = u_{n,i} = \mathrm{proj}(u_{n,i-1}, C(t_{n,i})) \,\varepsilon\, C(t_{n,i}) = C(t) ;$$

since $C(t)$ is a closed set, in the limit we get $u(t)\varepsilon\, C(t)$. Now, D is a dense subset of I; so, for any $t\varepsilon\, I$ a sequence $(t_n)\subset D$ can be found such that $t_n\to t$. Recalling that the multifunction C is Hausdorff-continuous and using the continuity of u we have then:

$$\mathrm{dist}(u(t),C(t)) = \lim_n \,\mathrm{dist}(u(t_n),C(t_n)) = 0 \;;$$

hence $u(t)\varepsilon\, C(t)$ for every t in I, as required by (2.3).

According to Theorem 2.5 we are now reduced to proving (2.10).

Let $z\varepsilon\, C(\tau)$ for every τ in $[s,t]$. If n is fixed, let us denote by j, $j+1,\ldots,k$ the values of i such that $s < t_{n,i} \le t$. By construction, $u_{n,i}$ is the projection of $u_{n,i-1}$ on $C(t_{n,i})$ and this set contains z; the property of projections implies that:

$$(19)\qquad\qquad (z-u_{n,i})\cdot(u_{n,i}-u_{n,i-1}) \ge 0.$$

On the other hand, a simple computation shows that in any real Hilbert space:

$$(20)\qquad\qquad x\cdot(x-y) \ge \tfrac{1}{2}\|x\|^2 - \tfrac{1}{2}\|y\|^2 .$$

We have

$$z\cdot[u_n(t)-u_n(s)] = z\cdot(u_{n,k}-u_{n,j-1}) = \sum_{i=j}^{k} z\cdot(u_{n,i}-u_{n,i-1}),$$

so that by (19) and (20):

$$(21)\quad z\cdot[u_n(t)-u_n(s)] \ge \sum_{i=j}^{k} u_{n,i}\cdot(u_{n,i}-u_{n,i-1})$$

$$\ge \sum_{i=j}^{k}\left(\tfrac{1}{2}\|u_{n,i}\|^2 - \tfrac{1}{2}\|u_{n,i-1}\|^2\right) = \tfrac{1}{2}\|u_{n,k}\|^2 - \tfrac{1}{2}\|u_{n,j-1}\|^2$$

$$= \tfrac{1}{2}\|u_n(t)\|^2 - \tfrac{1}{2}\|u_n(s)\|^2.$$

Taking (strong) limits as $n\to\infty$ we finally obtain (2.10):

$$z\cdot[u(t)-u(s)] \ge \tfrac{1}{2}\|u(t)\|^2 - \tfrac{1}{2}\|u(s)\|^2 . \qquad\qquad \square$$

Remark 3.6. Let us suppose that the multifunction C is defined in a noncompact interval $I = [0, +\infty[$, that C is Hausdorff-continuous and its values are closed convex subsets with nonempty interior in H. If u_0 is given in $C(0)$ then, for every n, Theorem 3.1 ensures the existence of exactly one solution to the sweeping process, defined in the interval $[0, n]$ and with initial value u_0; let us denote it by u^n. Uniqueness allows the following non-ambiguous definition, for $t \ge 0$:

$$u(t) := u^n(t) \qquad (\text{any } n \geq t).$$

This function u is continuous and has locally bounded variation and it clearly is the solution to the sweeping process in $[0, +\infty[$ (conveniently reformulated: du and $|du|$ are now σ-finite measures).

If, as in [Mor 9] (Prop.5.b, p.30), we assume that eventually the moving convex set contains a fixed ball:

$$t \geq T_1 \Rightarrow C(t) \supset \bar{B}(a, r) \, ,$$

then we easily deduce from (18) that $\text{var}(u; [T_1, +\infty[) \leq l(r, \| u(T_1) - a \|)$ and hence that u has bounded total variation. Therefore there exists a strong limit as $t \to +\infty$:

$$u_\infty := \lim_{t \to +\infty} u(t) \, .$$

It can be shown [Mor 9] that u_∞ belongs to the closure of the set of points that C eventually contains, i.e., to the closure of

$$\bigcup_{T \geq 0} \, \bigcap_{t \geq T} \, C(t) \, .$$

In the next theorem we study the behaviour of solutions with respect to changes in the initial values and in the considered multifunctions.

Theorem 3.7. *Let C and $\tilde{C}$ be two Hausdorff-continuous multifunctions from $I = [0, T]$ to closed convex sets having nonempty interior in a Hilbert space H. Let $u_0 \varepsilon C(0)$ and $\tilde{u}_0 \varepsilon \tilde{C}(0)$. Let u (respectively $\tilde{u}$) be the cbv solution to the sweeping process by C (resp. by $\tilde{C}$) with initial value u_0 (resp. $\tilde{u}_0$). We assume that for every t in I:*

(22) $$C(t) \supset \bar{B}(a, r) \, , \quad \tilde{C}(t) \supset \bar{B}(\tilde{a}, \tilde{r}) \, .$$

We define:

(23) $$\mu(t) := \sup_{0 < s \leq t} h(C(s), \tilde{C}(s)) \, .$$

Then:

(24) $$\| u(t) - \tilde{u}(t) \|^2 \leq \| u_0 - \tilde{u}_0 \|^2 + \mu(t) \left[\tfrac{1}{r} \| u_0 - a \|^2 + \tfrac{1}{\tilde{r}} \| \tilde{u}_0 - \tilde{a} \|^2 \right] \, .$$

Proof. As the interval may be shrunk arbitrarily, it is enough to consider the case $t = T$. Defining $u_{n,i}$ by (3.3)-(3.4) and $\tilde{u}_{n,i}$ in a similar manner, we already know that $u(T) = \lim u_{n,n}$ and $\tilde{u}(T) = \lim \tilde{u}_{n,n}$.

By Moreau's inequality on projections (Prop. 0.4.7), we have for $1 \leq i \leq n$:

$$\| u_{n,i} - \tilde{u}_{n,i} \|^2 \leq \| \operatorname{proj}(u_{n,i-1}, C(t_{n,i})) - \operatorname{proj}(\tilde{u}_{n,i-1}, \tilde{C}(t_{n,i})) \|^2$$

$$\leq \| u_{n,i-1} - \tilde{u}_{n,i-1} \|^2$$

$$+ 2\, (\| u_{n,i} - u_{n,i-1} \| + \| \tilde{u}_{n,i} - \tilde{u}_{n,i-1} \|)\, h(C(t_{n,i}), \tilde{C}(t_{n,i}));$$

thus

$$\| u_{n,i} - \tilde{u}_{n,i} \|^2 \leq \| u_{n,i-1} - \tilde{u}_{n,i-1} \|^2$$

$$+ 2\, \mu(T)\, (\| u_{n,i} - u_{n,i-1} \| + \| \tilde{u}_{n,i} - \tilde{u}_{n,i-1} \|).$$

Adding up these inequalities from $i = 1$ to n, we obtain after simplification:

$$(25) \quad \| u_{n,n} - \tilde{u}_{n,n} \|^2 \leq \| u_{n,0} - \tilde{u}_{n,0} \|^2$$

$$+ 2\, \mu(T) \sum_{i=1}^{n} (\| u_{n,i} - u_{n,i-1} \| + \| \tilde{u}_{n,i} - \tilde{u}_{n,i-1} \|)$$

$$\leq \| u_0 - \tilde{u}_0 \|^2 + 2\, \mu(T) [\operatorname{var}(u_n; 0, T) + \operatorname{var}(\tilde{u}_n; 0, T)]$$

$$\leq \| u_0 - \tilde{u}_0 \|^2 + 2\, \mu(T) [M + \tilde{M}],$$

where M is given by (13) and similarly $\tilde{M} = \| \tilde{u}_0 - \tilde{a} \| / (2\,\tilde{r})$; this is a consequence of hypothesis (22). Passing to the limit in (25) immediately produces (24) for the case $t = T$. $\square$

This section ends with a random or **parametrized version of the sweeping process**. This type of problem was studied for example by Castaing [Cas 2-5].

Let $I = [0, T]$ and H be a *separable* Hilbert space. Consider a measurable space $(\Omega, \mathcal{A})$ and a multifunction

$$C: I \times \Omega \to H$$

with closed convex values having nonempty interior in H. We assume that C satisfies the following conditions:

(i) *For every* $\omega \in \Omega$, *the multifunction* $t \to C(t, \omega)$ *is Hausdorff-continuous on* I;

(ii) *For every* $t \in I$, *the multifunction* $\omega \to C(t, \omega)$ *is* $\mathcal{A}$-*measurable* (see for instance [Cas-Val], Def. III-10).

We denote by $\mathfrak{B}(S)$ the σ-field of the borelian subsets of a topological space S.

The initial condition is now a $(\mathcal{A}, \mathcal{B}(H))$-*measurable function*

$$u_0 : \Omega \to H \,,$$

such that

(*iii*) *For every* $\omega \varepsilon \Omega$, $u_0(\omega) \varepsilon C(0, \omega)$.

In other words, u_0 is a measurable selection of $\omega \to C(0, \omega)$.

Theorem 3.8. *Under the hypotheses* (*i*)-(*iii*), *there exists one and only one* $(\mathcal{B}(I) \otimes \mathcal{A}, \mathcal{B}(H))$-*measurable function* $u: I \times \Omega \to H$ *such that, for every* $\omega \varepsilon \Omega$ *the function* $u_\omega(t) := u(t, \omega)$ *is cbv and:*

$$(26) \qquad u(0, \omega) = u_0(\omega) \qquad\qquad (\omega \varepsilon \Omega) \,;$$

$$(27) \qquad u(t, \omega) \varepsilon C(t, \omega) \qquad\qquad ((t, \omega) \varepsilon I \times \Omega) \,;$$

$$(28) \qquad -\frac{du_\omega}{|du_\omega|}(t) \varepsilon N_{C(t, \omega)}(u(t, \omega)) \qquad (\omega \varepsilon \Omega, \ |du_\omega|\text{-a. e. } t \varepsilon I).$$

Proof. Using the partition (1)-(2) of the interval I, for every n we define the following function $u_n : I \times \Omega \to H$:

$$(29) \qquad u_n(t, \omega) = \begin{cases} u_0(\omega) & \text{if } t \varepsilon I_{n,0} \\[2mm] \text{proj}\,(u_n(t_{n,i-1}, \omega), \, C(t_{n,i}, \omega)) & \text{if } t \varepsilon I_{n,i}, \ i \geq 1. \end{cases}$$

If ω is fixed, $t \to u_n(t, \omega)$ is a right-continuous step-function. On the other hand, if $t \varepsilon I_{n,0}$, then $\omega \to u_n(t, \omega) = u_0(\omega)$ is a $(\mathcal{A}, \mathcal{B}(H))$-measurable function, by hypothesis (iii); and if, by induction, this is assumed to be true for $t \varepsilon I_{n,i-1}$, then it will also be true for t in $I_{n,i}$, because the measurability of $u_n(t_{n,i-1}, \cdot)$ and of $C(t_{n,i}, \cdot)$ (see (ii)) implies measurability of $\text{proj}(u_n(t_{n,i-1}, \cdot), C(t_{n,i}, \cdot))$, by [Cas-Val] Theorem III-30. In short, u_n is separately right-continuous in t and measurable in ω. It follows that u_n is a $(\mathcal{B}(I) \otimes \mathcal{A}, \mathcal{B}(H))$-measurable function $-$ see e.g. [Cas 4] p.15.9 or [Cas-Val] Lemma III-14 (which applies here because u_n is "piecewise continuous" in t). We fix ω in Ω. Comparing the definition of u_n with (3)-(5) and applying Theorem 3.1, thanks to assumption (i), we obtain that:

$$\lim_n u_n(t, \omega) := u_\omega(t)$$

uniformly on I, where $u_\omega : I \to H$ is a cbv function, unique solution to the sweeping process by $t \to C(t, \omega)$ with initial value $u_0(\omega)$. In other words:

$$(30) \qquad u_\omega(0) = u_0(\omega) \,;$$

(31) $$u_\omega(t)\,\varepsilon\, C(t,\omega) \qquad (t\,\varepsilon\, I)\;;$$

(32) $$-\frac{du_\omega}{|\,du_\omega\,|}(t)\,\varepsilon\, N_{C(t,\omega)}(u_\omega(t)) \qquad (\,|\,du_\omega\,|\text{-}a.\ e.\ t\,\varepsilon\, I)\;.$$

The function $u(t,\omega):= u_\omega(t)$ is the pointwise limit of $(\mathfrak{B}(I)\otimes\mathcal{A},\ \mathfrak{B}(H))$-measurable functions (u_n); hence it is also measurable and it clearly satisfies (26)-(28).

Uniqueness is a consequence of the uniqueness of u_ω, for every ω. $\square$

2.4. Lower semicontinuous convex set in finite dimension

Let E be an Euclidian space, that is, a finite dimensional Hilbert space. Recall that a multifunction C from $I = [0,\,T]$ to E is **lower semicontinuous** (in short, lsc), respectively right lower semicontinuous, at the point $t_0\,\varepsilon\, I$ if, for every open set $U\subset E$ for which $C(t_0)\cap U\neq\emptyset$, there exists some $\epsilon>0$ such that $C(t)\cap U\neq\emptyset$ for every t in $[t_0-\epsilon,\,t_0+\epsilon]\cap I$, respectively in $[t_0,\,t_0+\epsilon]\cap I$. If this is true at every point of I, then we say that C is lsc, respectively **right lsc** on I. These definitions are easily extended to multifunctions defined on arbitrary topological spaces (see [Ber]).

The main result of this section is the following existence theorem.

Theorem 4.1. *Let C be a multifunction defined on I with closed convex values having nonempty interior in E. Let $u_0\,\varepsilon\, C(0)$. We assume that C satisfies one of the following conditions (A) or (B):*

(A)   *C is right lsc on I and there exist $a\,\varepsilon\, E$ and $r>0$ such that:*

(1) $$C(t)\supset \overline{B}(a,r) \qquad (t\,\varepsilon\, I)\,.$$

(B)   *C is lsc on I.*

Then the sweeping process (2.2)-(2.4) has one and only one rcbv solution $u\colon I\to E$. Moreover, under assumption (1), the estimate (3.18) of the total variation of the solution holds.

The proof is again based on the catching-up algorithm and on some preliminary results. The first of these is of geometrical nature and is related to

Lemma 2.3; in the literature we may find similar results, e. g. [Rol], Lemmas 4 and 5.

Lemma 4.2. *Let C be a multifunction with (closed) convex values having nonempty interior in an Euclidian space E and assume that C is lsc at a point t of some topological space. Let $\overline{B}(z, R)$ be a closed ball contained in the interior of $C(t)$. Then there exists a neighbourhood V of t such that:*

$$(2) \qquad\qquad C(\tau) \supset \overline{B}(z, R) \quad (\tau \varepsilon V) \; .$$

Proof. Let r' be such that $R < r' < \mathrm{dist}(z, E \backslash C(t))$ and take $\epsilon \varepsilon \;]0, r'-R[$. The compact set $\overline{B}(z, r')$ can be covered by a finite number of open balls $B(x_i, \epsilon/2)$ with $x_i \varepsilon \overline{B}(z, r')$, $i = 1, \ldots, N$. Since $x_i \varepsilon C(t) \cap B(x_i, \epsilon/2)$ and by definition of lower semicontinuity, we may then consider a neighbourhood V of t such that $C(\tau) \cap B(x_i, \epsilon/2) \neq \emptyset$, for every τ in V and for every i. Then, if $x \varepsilon B(x_i, \epsilon/2)$ we have $\mathrm{dist}(x, C(\tau)) < \epsilon$; thus,

$$e(\overline{B}(z, r'), C(\tau)) \leq e(\bigcup_{i=1}^{N} B(x_i, \epsilon/2), C(\tau)) \leq \epsilon \; .$$

Since $\overline{B}(z, r')$ and $C(\tau)$ are convex sets with nonempty interior, we may apply an inequality by Moreau (see Prop. 0.4.5), obtaining in this case:

$$r' = \mathrm{dist}(z, E \backslash \overline{B}(z, r')) \leq \mathrm{dist}(z, E \backslash C(\tau)) + e(\overline{B}(z, r'), C(\tau)) \; .$$

Thus, if $\tau \varepsilon V$, we have $\mathrm{dist}(z, E \backslash C(\tau)) \geq r' - \epsilon > R$ and (2) is proved. □

Corollary 4.3. *If C satisfies the hypotheses of Theorem 4.1, then C has the property (a) of Lemma 2.3.*

Proof. Given $t \varepsilon [0, T[$ and $z \varepsilon \mathrm{int}\, C(t)$, we may consider a closed ball $\overline{B}(z, R) \subset \mathrm{int}\, C(t)$. By hypothesis, C is a lsc multifunction on the interval I with the usual topology of $\mathbb{R}$ (case (B)) or with the right topology (case (A)). Lemma 4.2 ensures the existence a neighbourhood V of t in I satisfying (2). Since in either case this neighbourhood must contain an interval $[t, t+\epsilon]$ $(\epsilon > 0)$, (2.7) follows. □

Lemma 4.4. *Let y belong to a convex subset of E with nonempty interior C' and let $\epsilon > 0$. Then, there exist $x' \varepsilon C'$ and $r' > 0$ such that:*

$$(3) \qquad\qquad \overline{B}(x', r') \subset \mathrm{int}\, C' ,$$

$$(4) \qquad\qquad \frac{1}{2r'} \| y - x' \|^2 \leq \epsilon .$$

Proof. If y is an interior point of C', then take $x' = y$ and $r' < \operatorname{dist}(y, E \backslash C')$.

If y belongs to the boundary of C', then take any closed ball $\overline{B}(a, r)$ contained in the interior of C' and consider the points $a_\delta := y + \delta(a - y)$ with $0 < \delta < 1$. We have, by convexity of C':

$$\overline{B}(a_\delta, \delta r) = (1 - \delta) y + \delta \, \overline{B}(a, r) \subset \operatorname{int} C'.$$

It suffices to take $x' = a_\delta$ and $r' = \delta r$, with δ so small that

$$\frac{1}{2(\delta r)} \| y - a_\delta \|^2 = \frac{\delta}{2r} \| a - y \|^2 < \epsilon . \qquad\qquad \square$$

As in the case of a continuous moving convex set, we shall obtain the solution as the limit of right-continuous step-functions, constructed by a discretization of "time" t. Since in general the solution presents some discontinuities at points which we do not know *a priori*, it is crucial that the approximants be defined for every possible subdivision of the interval $I = [0, T]$. These subdivisions have the form:

$$P = \{ t_{P,i} : 0 \leq i \leq n(P) \} , \quad \text{with} \quad t_{P,0} = 0 < t_{P,1} < ... < t_{P,n(P)} = T.$$

In the set $\mathcal{P}$ of such subdivisions we consider the usual order: $P \geq P'$ if and only if $P \supset P'$, that is, if all the nodes of P' are also nodes of P. To each $P \varepsilon \mathcal{P}$ we associate a finite sequence $(u_{P,i})$ of elements of E, which are recursively defined by the formulas:

$$(5) \qquad\qquad\qquad u_{P,0} = u_0 ,$$

$$(6) \qquad\qquad u_{P,i} = \operatorname{proj}(u_{P,i-1}, C(t_{P,i})) \qquad (1 \leq i \leq n(P)) .$$

Then, we define a right-continuous step-function u_P:

$$(7) \qquad u_P(t) := \begin{cases} u_{P,i} & \text{if } t \varepsilon [t_{P,i}, t_{P,i+1}[\text{ and } 0 \leq i < n(P) \\[2mm] u_{P,n(P)} & \text{if } t = T. \end{cases}$$

We shall prove the following:

Theorem 4.5. *If C satisfies assumption (A) of theorem 4.1, then the generalized sequence (or net) $(u_P)_{P \varepsilon \mathcal{P}}$ converges pointwisely to the solution to the sweeping process by C with initial value u_0. Furthermore, this solution satisfies the estimate (3.18).*

Proof. By hypothesis (1), we have $C(t_{P,i}) \supset \bar{B}(a,r)$ for every P and i. Then, Lemma 0.4.4 and the definition of u_P ensure that

$$(8) \quad \mathrm{var}(u_P; I) = \sum_{i=1}^{n(P)} \| u_{P,i} - u_{P,i-1} \| \leq l(r, \| u_0 - a \|) := M \leq \frac{1}{2r} \| u_0 - a \|^2 ,$$

$$(9) \quad\quad\quad\quad\quad\quad\quad\quad u_P(t) = u_0 .$$

Hence,

$$\| u_P(t) \| \leq \| u_0 \| + M := L .$$

Thus the generalized sequence (u_P) is bounded in the uniform norm and in total variation. Applying Theorem 0.2.2, we know that there is a filter $\mathcal{F}$ finer than the filter of sections of $\mathcal{P}$ and that there is a function of bounded variation $u : I \to E$ which is the weak pointwise limit of (u_P) with respect to $\mathcal{F}$. In other words, and since E is finite-dimensional, u is a strong pointwise generalized sublimit of (u_P):

$$(10) \quad\quad\quad \forall t \varepsilon I: \quad \lim_{\mathcal{F}} \| u_P(t) - u(t) \| = 0 .$$

Also, by the same theorem, 0.2(9), we have $\mathrm{var}(u; I) \leq M$, that is (3.18).

We now prove that *u is the solution to the sweeping process* (2.2)-(2.4), *unique* by the classical monotonicity argument.

1°)   From (9) and (10) it is clear that $u(0) = u_0$, i. e. (2.2).

2°)   On the other hand, for every $t \varepsilon]0, T]$, whenever P belongs to the section determined by the subdivision $P_0 = \{0,\ t,\ T\}$, that is, if $P \supset P_0$, then t is one of the nodes in P and by definition of u_P (see (6) and(7)) it follows that $u_P(t) \varepsilon C(t)$. Now, (10) and closedness of $C(t)$ imply that:

$$u(t) \varepsilon C(t) .$$

This being obvious for $t = 0$, because of 1°), we have fully established (2.3).

3°)   We prove that *u* is right-continuous:

$$(11) \quad\quad\quad u^+(t) = u(t) \quad\quad (\ t \varepsilon [0, T[\ ) .$$

For fixed t and arbitrary $\epsilon > 0$, since $u(t)$ belongs to the convex set with nonempty interior $C(t)$ we may take some $x' \varepsilon C(t)$ and some $r' > 0$ such that $\bar{B}(x', r') \subset \mathrm{int}\, C(t)$ and $\| u(t) - x' \|^2 /(2r') < \epsilon$ (by Lemma 0.4.4). Thanks to (10), there is then some $F \varepsilon \mathcal{F}$ such that, for every $P \varepsilon F$, both $t \varepsilon P$ and $\| u_P(t) - x' \|^2 /(2r') < \epsilon$ are true.

By Lemma 4.2 (considering I with the right topology) there is some positive δ for which $t+\delta \leq T$ and $C(\tau) \supset \bar{B}(x',r')$ whenever $\tau \varepsilon [t, t+\delta]$. Thus, if we take $x_0 = u_P(t) = u_{P,i}$ (for some i, by definition of u_P) as the initial value of the finite sequence in Lemma 0.4.4, we obtain the estimate:

$$(11') \qquad \operatorname{var}(u_P;]t, t+\delta]) \leq \frac{1}{2r'}\| u_P(t) - x'\|^2 < \epsilon \qquad (P \varepsilon F) .$$

In particular, for every $\tau \varepsilon [t, t+\delta]$: $\| u_P(\tau) - u_P(t) \| < \epsilon$. Hence, taking limits along $\mathcal{F}$ using (10), it follows that $\| u(\tau) - u(t) \| \leq \epsilon$ and the proof of (11) is complete.

4°)   We show that:

$$(12) \qquad \forall t \varepsilon\,]0, T]: \ u(t) = \operatorname{proj}(u^-(t), C(t)) ,$$

which if combined with (11) ensures that:

$$-\frac{du}{|du|}(t) = \frac{u^-(t) - \operatorname{proj}(u^-(t), C(t))}{\| u^-(t) - \operatorname{proj}(u^-(t), C(t)) \|} \ \varepsilon \ N_{C(t)}(u(t)) ,$$

by a property of projections and similarly to section 1.4.

Let $t \varepsilon\,]0, T]$ and $\epsilon > 0$. For every $x \varepsilon E$, we denote by $D(x)$ the closed convex hull of $\bar{B}(a, r) \cup \{x\}$. Notice that, by assumption (1), if x belongs to $C(t)$ then $D(x) \subset C(t)$.

Let $x' \varepsilon D(u^-(t))$ and $r' > 0$ be such that $\bar{B}(x', r') \subset \operatorname{int} D(u^-(t))$ and $\| u^-(t) - x'\|^2/(2r') < \epsilon/3$ (by Lemma 4.4). There is $t' < t$ for which $\| u(t') - u^-(t) \| < \epsilon/3$ and $\| u(t') - x'\|^2/(2r') < \epsilon/3$, by definition of left-limit. Moreover, the multifunction $x \to D(x)$ is clearly Hausdorff-continuous and *a fortiori* lower semicontinuous. Thus Lemma 4.2 allows us to choose t' in such a way that also:

$$(13) \qquad \bar{B}(x', r') \subset D(u(\tau)) \subset C(\tau) \qquad (\tau \varepsilon [t', t[ ) .$$

On the other hand, pointwise convergence (10) implies the existence of some $F \varepsilon \mathcal{F}$ such that, for all $P \varepsilon F$, we have $t, t' \varepsilon P$, $\| u_P(t') - u(t') \| < \epsilon/3$ and $\frac{1}{2r'}\| u_P(t') - x'\|^2 < \epsilon/3$. By Lemma 0.4.4, from (13) and the last inequality we deduce that, for any such τ:

$$(14) \qquad \| u_P(\tau) - u^-(t') \| \leq \operatorname{var}(u_P;[t', t[) < \tfrac{\epsilon}{3} ;$$

whence:

$$(15) \qquad \| u_P(\tau) - u^-(t) \| < \tfrac{\epsilon}{3} + \| u_P(t') - u(t') \| + \| u(t') - u^-(t) \| < \epsilon$$

If τ is the greatest $t_{P,i}$ contained in the interval $[t', t[$, then by definition

$u_P(t) = \mathrm{proj}(u_P(\tau), C(t))$ and we can write, for $P \varepsilon F$:

$$\| u_P(t) - u^-(t) \| \leq \| \mathrm{proj}(u_P(\tau), C(t)) - \mathrm{proj}(u^-(t), C(t)) \|$$

$$+ \| \mathrm{proj}(u^-(t), C(t)) - u^-(t) \|$$

$$\leq \| u_P(\tau) - u^-(t) \| + \mathrm{dist}(u^-(t), C(t)) ,$$

by the nonexpansiveness of projections. Thus, using (10) and (15):

$$\| u(t) - u^-(t) \| \leq \mathrm{dist}(u^-(t), C(t)) + \epsilon .$$

Since $u(t) \varepsilon C(t)$ and ϵ is arbitrary, this means that $u(t)$ actually is the projection of $u^-(t)$ into $C(t)$.

5°)   To end the proof, we show that the differential inclusion (2.4) is satisfied at $|du|$-almost every continuity point of u. Thanks to the preceding properties of u, to Corollary 4.3 and to Lemma 2.4 we only need to verify that **condition (2.10)** holds. This is done, *mutatis mutandis*, as in the proof of Lemma 3.5 (see (3.19)-(3.21)).

If $z\varepsilon C(\tau)$ for every $\tau\varepsilon[s,t]$ then, for any partition P, there exist j and k such that:

$$(16) \qquad z.(u_P(t) - u_P(s)) = \sum_{i=j}^{k} z.(u_{P,i} - u_{P,i-1}) \geq \sum_{i=j}^{k} u_{P,i}.(u_{P,i} - u_{P,i-1})$$

$$\geq \tfrac{1}{2} \| u_P(t) \|^2 - \tfrac{1}{2} \| u_P(s) \|^2 ,$$

where we use (6) and a property of projections. Taking limits along $\mathcal{F}$, we finally obtain (2.10).

Recall that we began by considering an arbitrary pointwise convergent (generalized) subsequence of (u_P) and we have just shown that its limit is necessarily the unique solution u to the sweeping process. Since the given (generalized) sequence is relatively compact for the topology of pointwise convergence, it follows that (u_P) itself converges to u:

$$(17) \qquad\qquad u(t) = \lim_{\mathcal{P}} u_P(t)$$

and the theorem is completely established. $\square$

Remark 4.6. We assume now that C is only **right lower semicontinuous**. By Lemma 4.2, any closed ball contained in the interior of $C(0)$ will remain in the interior of $C(\tau)$ for τ in some neighbourhood $I_0 = [0, \delta]$ of $t = 0$. Then, the preceding theorem may be applied to the restriction of the multifunction C to

I_0, thus ensuring the existence in that interval of a solution to the sweeping process, obviously called a **local solution**.

Under **assumption (B)**, that is, if C is **lower semicontinuous**, we can paste together a finite number of local solutions and obtain a global solution (defined on the preassigned interval $[0, T]$). Let us see how this can be done.

For each $t \varepsilon I$, we choose by Lemma 4.2 an interval I_t which is an open neighbourhood of t in I and is such that every $C(s)$ with $s \varepsilon I_t$ contains some fixed ball. By compactness of I, a finite number of these intervals covers I. By ordering their respective endpoints, we find $s_0 < s_1 < ... < s_p = T$ such that in every interval $J_k = [s_{k-1}, s_k]$ we have $C(\tau) \supset \overline{B}(a_k, r_k)$. In each J_k, the multifunction C satisfies assumption (A) of Theorem 4.1. Thus, by Theorem 4.5, in $J_1 = [0, s_1]$ the sweeping problem has the rcbv solution u_1, with $u_1(0) = u_0$. Next, since $u_1(s_1) \varepsilon C(s_1)$ we can solve the sweeping problem on J_2 having that initial value at $t = s_1$ (that zero is not the left endpoint of the interval where the solution is to be found is of course irrelevant). We denote by u_2 the rcbv solution with $u_2(s_1) = u_1(s_1)$. Repeating this argument, we define the solutions $u_1, u_2, ..., u_p$ on the respective intervals, the "initial" conditions being $u_{k+1}(s_k) = u_k(s_k)$.

The solution to the sweeping process on the whole of I is the "pasted" or "continued" function u, defined by

$$(18) \qquad u(t) = u_k(t) , \quad \text{if } t \varepsilon J_k \ (1 \leq k \leq p).$$

In fact, u is clearly a right-continuous function of bounded variation that satisfies (2.2) $u(0) = u_0$ and (2.3) $u(t) \varepsilon C(t)$ $(t \varepsilon I)$. Furthermore, by construction, for each k,

$$-\frac{du_k}{|du_k|}(t) \ \varepsilon \ N_{C(t)}(u_k(t))$$

at $|du_k|$-almost every t in J_k; and we have $du = du_k$ in the (relative) interior of these intervals. It follows that the differential inclusion (2.4) is satisfied at $|du|$-almost every point of $I \setminus \{s_1, ..., s_{p-1}\}$. Those of the remaining nodes where u is continuous have zero measure, hence we only need to study the case of a discontinuity occurring at some s_k. As $u^+(s_k) = u_{k+1}(s_k) = u_k(s_k)$ and $u^-(s_k) = u_k^-(s_k)$, we have:

$$(19) \qquad -\frac{du}{|du|}(s_k) = -\frac{u_k(s_k) - u_k^-(s_k)}{\| u_k(s_k) - u_k^-(s_k) \|} = -\frac{du_k}{|du_k|}(s_k),$$

hence:

$$-\frac{du}{|\,du\,|}\,(s_k)\,\varepsilon\,N_{C(s_k)}\,(u_k(s_k)) = N_{C(s_k)}\,(u(s_k))\,,$$

thus completing the **proof of Theorem 4.1**.   ☐

Remark 4.7. If a multifunction C defined on the **noncompact interval** $I=[0,+\infty[$ is otherwise under the conditions of the theorem, then we may still ensure the existence of a unique solution to the sweeping process. This is the right-continuous function with locally bounded variation defined as in Remark 3.6 or by an obvious version of the "pasting method" explained above.

Remark 4.8. If C not only satisfies the assumptions of Theorem 4.1 but is also **left upper semicontinuous** (or, to be precise, if its graph is closed in the topological product of I, with its left-topology, and E) then the solution to the sweeping process is **continuous**. We only need to show that, in this case, u is also left-continuous. Let $t\varepsilon\,]0,\,T]$ and take $t_n\uparrow t$; as $(t_n,u(t_n))\,\varepsilon$ graph C, their limit in the afore-mentioned product topology, which is $(t,u^-(t))$, must also belong to the graph of C. That is, $u^-(t)\,\varepsilon\,C(t)$ and by (12) $u(t)=u^-(t)$.

Remark 4.9. In Theorem 4.5, the use of the **generalized sequence of approximants** (u_P) is essential. In fact, consider instead a sequence (u_n) of step-functions, defined by (3.3)-(3.5), whose discretization nodes $t_{n,i}$ are given by (3.1) or any other law with diminishing "step" (e. g. $t_{n,i}=iT/2^n$). We construct a simple multifunction for which this discretization procedure is useless. Denoting by N the countable set of all the nodes of all the would-be approximants, there is always some $t^*\notin N$. We define

$$(20)\qquad\qquad C(t)=\begin{cases}[0,3] & \text{if } 0\le t<t^* \\[2pt] [2,3] & \text{if } t=t^* \\[2pt] [1,3] & \text{if } t^*<t\le T\end{cases}\quad;$$

this is a lower semicontinuous multifunction with closed convex values having nonempty interior in $E=\mathbb{R}$. Choose the initial value $u_0=0$. If $t_{n,i}$ is the smallest node greater than t^*, then $u_n(t)=0$ if $t<t_{n,i}$ and $u_n(t)=1$, otherwise. Hence the pointwise limit of the sequence (u_n) is the function v, given by $v(t)=0$ (if $t\le t^*$) and $v(t)=1$ (if $t>t^*$), while the solution is $u(t)=0$ (if $t<t^*$) and $u(t)=2$ (if $t\ge t^*$). Note that even v^+ is not the solution.

Remark 4.10. The preceding example (20) and example 1.1.4 show the **ineffectiveness of the Yosida regularization approach** in the present case of a sweeping process by a lsc moving convex set. However, the Hausdorff-continuous case looks more promising.

Instead, we now turn our attention to strengthening the pointwise convergence in (17). As in the continuous case (Theorem 3.1) we prove that the approximants (u_P) **converge uniformly** to the solution to the lsc sweeping process. Hence, they also converge in the sense of filled-in graphs (Theorem 0.3.1).

Theorem 4.11. *If C is (right) lsc and satisfies* (1), *then the generalized sequence* (u_P) *of the step-approximants* (7) *converges uniformly to the solution to the sweeping process:*

$$(21) \qquad \lim_{\mathcal{P}} \| u_P - u \| = 0 .$$

Proof. First, we remark that it is really not that restrictive to assume that (1) holds, particularly when C is lsc from both sides, as previous arguments have shown.

Let $\epsilon > 0$ be given. To each $t \varepsilon I$ we shall associate a neighbourhood I_t of t in I and $P_t \varepsilon \mathcal{P}$ such that if $P \geq P_t$ (that is, if $P \supset P_t$) then:

$$(22) \qquad \| u_P(s) - u(s) \| < 3\epsilon \quad (s \varepsilon I_t) .$$

It is known that I can be covered by a finite number of such neighbourhoods; we denote by P_0 the union of the respective subdivisions P_t . By (22) it will then be clear that $\| u_P - u \|_\infty \leq 3\epsilon$ whenever $P \geq P_0$, proving (21).

To prove (22), let us fix $t \varepsilon]0, T[$ (the endpoints require easy "unilateral" versions of the reasonings that follow). Proceeding as in the proof of Theorem 4.5, 3°), we select a subdivision P_1 and $\delta > 0$ in such a way that (cf. (11') and use (17) in place of (10)):

$$\mathrm{var}(u_P; \,]t, t+\delta]) < \epsilon \qquad (P \geq P_1) .$$

In particular, for every $s \varepsilon I^* = \,]t, t+\delta]$:

$$(23) \qquad \| u_P(s) - u_P(t) \| < \epsilon .$$

Thus, taking limits with respect to P:

(24) $$\| u(s) - u(t) \| \leq \epsilon .$$

By (17), choose $P_2 \epsilon \mathcal{P}$ such that whenever $P \geq P_2$ we have:

(25) $$\| u_P(t) - u(t) \| \leq \epsilon .$$

Combining (23)-(25), we obtain:

(26) $$\| u_P(s) - u(s) \| < 3\epsilon ,$$

for all $s \epsilon I^*$ and every $P \geq P_1 \cup P_2$.

We now turn our attention to what happens to the left of t. As in the proof of Theorem 4.5, 4°) there is some $t' < t$ and some $P_3 \epsilon \mathcal{P}$ such that for $P \geq P_3$ and $\tau \epsilon I' := [t',t[$ there hold both (14) $\| u_P(\tau) - u_P(t') \| < \epsilon/3$ and (15) $\| u_P(\tau) - u^-(t) \| < \epsilon$. From (14) we get, upon passing to the limit:

(27) $$\| u(\tau) - u(t') \| \leq \tfrac{\epsilon}{3} ;$$

so, letting $\tau \uparrow t$:

(28) $$\| u^-(t) - u(t') \| \leq \tfrac{\epsilon}{3} .$$

Thus, by (15), (27) and (28):

(29) $$\| u_P(\tau) - u(\tau) \| \leq \tfrac{5\epsilon}{3} \quad (\tau \epsilon I' , P \geq P_3) .$$

Let $I_t := I' \cup \{t\} \cup I^* = [t',t+\delta]$. Thanks to (25), (26) and (29), the inequality (22) is true for every partition $P \geq P_t := P_1 \cup P_2 \cup P_3$. $\square$

The next set of results concerns the dependence of the solution to the sweeping process on the data, that is, on the moving convex set C and on the initial value u_0. The first one is similar to Theorem 3.7.

Theorem 4.12. *Let C and $\tilde{C}$ be two right lsc multifunctions defined on $I=[0,T]$ with closed convex values having nonempty interior in an Euclidian space E. Let $u_0 \epsilon C(0)$ and $\tilde{u}_0 \epsilon \tilde{C}(0)$. We assume that:*

(30) $$\forall t \epsilon I: \quad C(t) \supset \overline{B}(a,r) , \quad \tilde{C}(t) \supset \overline{B}(\tilde{a},\tilde{r}) .$$

We denote by u and $\tilde{u}$ the solutions to the sweeping processes by C and $\tilde{C}$ with initial values u_0 and $\tilde{u}_0$ respectively. Then, for every t:

(31) $$\| u(t) - \tilde{u}(t) \|^2 \leq \| u_0 - \tilde{u}_0 \|^2 + 2\mu(t) [l(r, \| u_0 - a \|) + l(\tilde{r}, \| \tilde{u}_0 - \tilde{a} \|)]$$

$$\leq \| u_0 - \tilde{u}_0 \|^2 + \mu(t) [\tfrac{1}{r} \| u_0 - a \|^2 + \tfrac{1}{\tilde{r}} \| \tilde{u}_0 - \tilde{a} \|^2] ,$$

where $\mu(t) := \sup \{ h(C(s), \tilde{C}(s)) : 0 < s \leq t \}$ and $\mu(0) = 0$.

Proof. Similar to that of Theorem 3.7. Without loss of generality, we consider $t = T$. We have, by (7) and (17), $u(T) = \lim_{\mathcal{P}} u_{P,n(P)}$ and $\tilde{u}(T) = \lim_{\mathcal{P}} \tilde{u}_{P,n(P)}$ (analogously defined). By Moreau's inequality on projections into two convex sets:

$$\| u_{P,i} - \tilde{u}_{P,i} \|^2 \leq \| u_{P,i-1} - \tilde{u}_{P,i-1} \|^2$$
$$+ 2\mu(t) \left(\| u_{P,i} - u_{P,i-1} \| + \| \tilde{u}_{P,i} - \tilde{u}_{P,i-1} \| \right) ;$$

summing up and simplifying:

$$\| u_{P,n(P)} - \tilde{u}_{P,n(P)} \|^2 \leq \| u_0 - \tilde{u}_0 \|^2 + 2\mu(t) \left[\mathrm{var}(u_P;I) + \mathrm{var}(\tilde{u}_P;I) \right].$$

These total variations are easily bounded above, thanks to assumption (30), Theorem 4.5 and (3.18). Taking limits we obtain (31) for $t = T$. $\qquad\square$

Remark 4.13. More generally, we have

$$(32) \qquad \| u(t) - \tilde{u}(t) \|^2 \leq \| u_0 - \tilde{u}_0 \|^2 + 2\mu(t) \left[M + \tilde{M} \right],$$

where M and $\tilde{M}$ are arbitrary upper bounds of the total variations of all the approximants (u_P) and $(\tilde{u}_P)$. But, at least by using this technique, we are not allowed to take $M = \mathrm{var}(u;I)$ and $\tilde{M} = \mathrm{var}(\tilde{u}; I)$ in (32). For instance, from the vague (i.e., weak-$*$) convergence of the measures du_P to du (cf. Theorem 0.2.2(ii))' it does not follow that $\mathrm{var}(u_P;I) = \int |du_P|$ converges to $\mathrm{var}(u;I) = \int |du|$ or equivalently that $\limsup \mathrm{var}(u_P;I) \leq \mathrm{var}(u;I)$.

Theorem 4.12 implies the following corollary, which states in a more direct way the continuous dependence of the solutions on the data:

Corollary 4.14. *Let C and C_n $(n \geq 1)$ be lsc multifunctions from $I = [0, T]$ to closed convex sets with nonempty interior in an Euclidian space E. Let $u_{n,0} \varepsilon\, C_n(0)$ $(n \geq 1)$ be such that $u_{n,0} \to u_0 \varepsilon\, C(0)$ and denote by u_n and u the solutions to the sweeping processes by C_n and C with initial values $u_{n,0}$ and u_0 respectively. Suppose that:*

$$(33) \qquad \lim_n \left[\sup_{t \varepsilon I} h(C_n(t), C(t)) \right] = 0 .$$

Then, (u_n) converges to u uniformly on I.

Proof. We have already seen that I can be partitioned into a finite number of intervals $I_i = [t_{i-1}, t_i]$ $(1 \leq i \leq p)$ such that, for some well chosen closed balls:

$$(34) \qquad \overline{B}(a_i, 2r) \subset C(t) \qquad (t \varepsilon I_i).$$

Writing $\mu_n = \sup\{h(C_n(t), C(t)): t\varepsilon I\}$ and using (33), we choose n_0 such that $\mu_n \leq r$ whenever $n \geq n_0$. If $t\varepsilon I_i$, then an inequality by Moreau (Prop. 0.4.5) ensures that:

$$\operatorname{dist}(a_i, E \backslash C_n(t)) \geq \operatorname{dist}(a_i, E \backslash C(t)) - e(C(t), C_n(t)) \geq 2r - \mu_n \geq r,$$

hence:

$$(35) \qquad \overline{B}(a_i, r) \subset C_n(t) \qquad (t\varepsilon I_i, \ n \geq n_0).$$

We prove that uniform convergence holds on the first subinterval $I_1 = [t_0, t_1] = [0, t_1]$. Given a positive ϵ, the hypotheses show that there exists $n_1 \geq n_0$ such that for every $n \geq n_1$:

$$(36) \qquad \| u_{n,0} - u_0 \| \leq \epsilon/\sqrt{2},$$

$$(37) \qquad \mu_n [\, 2 \| u_0 - a_1 \| + \epsilon/\sqrt{2}\,]^2 / r \leq \epsilon^2/2.$$

We apply Theorem 4.12 to C_n, C, $u_{n,0}$ and u_0, taking (34)–(37) into account, and we obtain from (31) the estimate:

$$\| u_n(t) - u(t) \|^2 \leq \| u_{n,0} - u_0 \|^2 + \mu_n [\tfrac{1}{r} \| u_{n,0} - a_1 \|^2 + \tfrac{1}{2r} \| u_0 - a_1 \|^2]$$

$$\leq \epsilon^2/2 + \mu_n [(\| u_0 - a_1 \| + \epsilon/\sqrt{2})^2 + \| u_0 - a_1 \|^2]/r \leq \epsilon^2,$$

that is,

$$(38) \qquad \| u_n(t) - u(t) \| \leq \epsilon \qquad (t\varepsilon I_1, \ n \geq n_1).$$

The uniform convergence of (u_n) to u on $I_2 = [t_1, t_2]$ is obtained in the same way because, by (38), the initial values at $t = t_1$ satisfy $u_n(t_1) \to u(t_1)$. Thus, we consecutively establish the uniform convergence on each subinterval, and hence on I. $\qquad\qquad\square$

Remark 4.15. The question of existence of a solution to the sweeping process by a lower semicontinuous moving convex set with nonempty interior in an **infinite-dimensional** Hilbert space arises naturally. Let us point out some of the difficulties inherent in such a generalization to arbitrary dimension.

a) Defining the net of approximants (u_p) exactly as before, we can now extract from it a subnet which converges pointwisely weakly to some bv function u (see 0.2(8)). Reasoning as in the proof of Theorem 4.5, we have $u(0) = u_0$ and $u(t)\varepsilon C(t)$ $(t\varepsilon I)$ but we are unable to obtain the right-continuity of u (11) or the condition (12) at jump points (strong pointwise convergence is needed).

Even at continuity points, the weak pointwise convergence does not allow us to infer (2.10) from (16).

b) Another likely method, which may be called the **inner approximation method,** encounters similar difficulties. Using Michael's selection theorem [Mic] and Castaing's representation theorem [Cas 6] (see also [Bres 1], [Fry]), we define an increasing sequence of Hausdorff-continuous multifunctions (C_n) with nonempty interior such that:

$$C(t) = \mathrm{cl}\Big(\bigcup_n C_n(t)\Big).$$

If for each n, u_n is the solution to the sweeping process by C_n such that $u_n(0) = u_0$ (Theorem 2.1), then (not without some work) we prove that there is a sublimit u of (u_n) that obviously satisfies (2.2) and (2.3), but apparently not (2.4) (the jump points being the main problem).

c) Recently, Valadier [Val 1] applied the inner approximation method to deduce the existence result for continuous sweeping processes (and even a little more, in the style of Remark 4.8) from the well-known existence for Lipschitz-continuous ones. In any case, his solution is continuous. More detailed information is presented in Chapter 5, §5.2.

Remark 4.16. If, under the assumptions of Theorem 4.1, we know that:

$$(39) \qquad\qquad \forall t \varepsilon\, I: \quad b \,\varepsilon\, C(t),$$

for some $b \varepsilon E$, then the solution to the sweeping process by C with initial value u_0 satisfies:

$$(40) \qquad\qquad \forall t \varepsilon\, I: \quad \| u(t) - b \| \leq \| u_0 - b \|.$$

In fact, by Lemma 0.4.4 and by definition of the approximants:

$$\| u_P(t) - b \| = \| u_{P,i} - b \| \text{ (for some } i) \leq \| u_{P,0} - b \| = \| u_0 - b \|$$

and we only need to use (17).

This result applies also to the infinite-dimensional case of sections 2 and 3. Moreover, with the same technique we may obtain:

$$(41) \qquad\qquad t \rightarrow \| u(t) - b \| \text{ is nonincreasing.}$$

Chapter 3

Inelastic Shocks with or without Friction: Existence Results

3.1. Introduction

In this Chapter, we prove convergence for algorithms arising in the study of the dynamics of a mechanical system with a finite number of degrees of freedom. It is assumed that the system is subjected to a unique unilateral constraint, with inelastic contact and possibly isotropic dry friction (Coulomb). The theoretical foundation is the formulation of Moreau (see [Mor 11-14] or [Jea-Mor]), which we briefly review.

To simplify, it is assumed that the system is represented by a point q in the n-dimensional Euclidian space E; in other words, by the use of local coordinates, the manifold of configurations is identified with E. The evolution (motion) of the system during a certain time interval I is described by a function $q: I \rightarrow E$, where, without loss of generality, we take $I=[0, T]$, $T>0$.

The system is subjected to a unique **unilateral constraint** geometrically expressed by an inequality $f(q) \leq 0$; that is, $q(t)$ must belong to the closed region

$$(1) \qquad L := \{ q \varepsilon E : f(q) \leq 0 \} ,$$

where $f: E \rightarrow \mathbb{R}$ is a C^1 function, not depending on time (the system is "scleronomic"), and whose gradient is never zero; the last requirement only needs to be met in a neighbourhood of the following hypersurface:

$$(2) \qquad S := \{ q \varepsilon E : f(q)=0 \} .$$

The case of several unilateral constraints is considered, but not solved, in section 4.4.

With this formalism, here somewhat simplified, we are able to study systems of rigid bodies which can contact ($f(q)=0$) or move apart ($f(q) < 0$) but not interpenetrate (see [Mor 12] , [Jea-Mor]); or systems of bodies linked by an inextensible string, the equality meaning in this case that the string is strained.

We single out, however, the following simple model: $q(t)$ is the position at time t of a material point, a small object of unit mass, confined to a region L of the physical space $(E=\mathbb{R}^3)$ bounded by the fixed material wall S. The object is submitted to the action of a force $p(t, q)$, depending on time and position (the dependence on velocity is also worth considering). The motion $t \to q(t)$ takes place in L and we assume that the right-velocity

$$v^+(t) := \dot{q}^+(t) := \lim_{h\downarrow 0} \frac{q(t+h)-q(t)}{h}$$

and the left-velocity $v^-(t)$ exist for every t.

An instant t when there is contact with S is a local maximum of the function $s \to f(q(s))$; hence, derivating from both right and left, we obtain:

$$(3) \qquad \nabla f(q(t)) . v^+(t) \leq 0,$$

$$(4) \qquad \nabla f(q(t)) . v^-(t) \geq 0 .$$

Condition (3) means that $v^+ = v^+(t)$ must belong to a halfspace, called the tangent halfspace to L at the point $q(t)$. On the other hand, if at time t the material point finds itself in the interior of the region L, i. e., if $f(q(t)) < 0$, then no restriction is imposed on the right-velocity v^+. To deal simultaneously with both cases, we introduce the set $V(q)$ of kinematically admissible right-velocities at the point $q \varepsilon L$ which we call the tangent cone to L at q. More precisely, we define $V(q)$ even outside of L, because while applying approximation algorithms the unilateral constraint may momentarily be violated. We put:

$$(5) \qquad V(q) = \begin{cases} \{w \varepsilon E : \ w . \nabla f(q) \leq 0\} & , \text{ if } f(q) \geq 0 \\ E & , \text{ if } f(q) < 0 . \end{cases}$$

Omitting t for notational simplicity, (3) and (4) are rewritten as:

$$(6) \qquad v^+ \varepsilon V(q) ;$$

$$(7) \qquad - v^- \varepsilon V(q) .$$

Observe that $V(q)$ is either the whole space E or a halfspace (the gradient being nonzero by hypothesis), hence its interior is nonempty:

$$(8) \qquad \text{int } V(q) \neq \emptyset .$$

On the other hand, it is easily verified that the multifunction $q \to V(q)$ is lower semicontinuous on E (Lemma 2.1). These two facts ensure that estimates and properties similar to those of Chapter 2 hold for the algorithm we shall use.

If an episode of smooth motion ends at the instant t with left-velocity $v^- \varepsilon V(q)$, that is, if $f(q)=0$ and $v^- . \nabla f(q) > 0$ then a **shock** necessarily occurs: right-velocity is different from left-velocity.

Let us consider first the case of a perfect **frictionless contact**. The material point (system, body) then experiences from the part of the "wall" S a normal reaction force r, pointing inwards to L at the contact point q. In other words, r is such that ([Mor 11] (2.2)-(2.3) ; [Mor 12] (3.7)-(3.8)) :

$$(9) \qquad \exists \lambda \varepsilon \mathbb{R} : \quad r = -\lambda \nabla f(q) ;$$

$$(10) \qquad \lambda \geq 0 .$$

This inequality expresses the absence of adhesion to S. By "integrating" the reaction (or liaison) force r "for the (infinitesimally small) duration of the shock" and assuming that r does not change direction, we obtain the *liaison percussion* P. Thanks to (9) and (10), P satisfies ([Mor 12] (5.1), (5.2)):

$$(11) \qquad P = v^+ - v^- = -\alpha \nabla f(q) \quad (\alpha > 0) .$$

Conditions (6) and (11) alone do not univocally determine v^+, the velocity after the shock, once the velocity before the shock v^- is known. We say that it is an **elastic shock** if the (kinetic) energy is conserved; that is, if:

$$(12) \qquad \| v^+ \| = \| v^- \| ,$$

in the euclidian norm of E, which is chosen so that the kinetic energy is given by

$$(13) \qquad E(v) = \tfrac{1}{2} \| v \|^2 .$$

Taken together, (11) and (12) imply that v^+ is obtained from v^- by geometrical reflection with respect to the tangent hyperplane to L at point q.

The other ideal extreme case, which is the object of this study, is that of a **soft** or **inelastic shock**: the velocity after the shock is tangential, i. e.,

$$(14) \qquad v^+ . \nabla f(q) = 0 .$$

Accounting for (11), it is geometrically clear that v^+ is the orthogonal projection of v^- into the tangent hyperplane

$$(15) \qquad T(q) = \{ w: \ w . \nabla f(q) = 0 \} ;$$

in particular,

$$(16) \qquad \| v^+ \| < \| v^- \| ,$$

implying that the shock is **dissipative**. Also

$$(17) \qquad\qquad v^+ = \mathrm{proj}\,(v^-,\ V(q))\ ,$$

the proximal point or projection of v^- into the tangent halfspace $V(q)$. This is a "principle of economy": among the kinematically admissible right-velocities, the nearest one is chosen.

By elementary Convex Analysis or simply by looking at (11), we infer that, equivalently:

$$(18) \qquad\qquad v^- - v^+ \ \varepsilon\ N_{V(q)}(v^+)\ ,$$

the outward normal cone to $V(q)$ at the point v^+, which in this case is simply the outward normal halfline, spanned by $\nabla f(q)$.

We turn our attention to episodes of smooth motion. If the acceleration $\ddot q = \dot v$ is well defined, then Lagrange's equation holds:

$$(19) \qquad\qquad \ddot q = p(t, q) + r\ ,$$

which leads to the following differential inclusion:

$$(20) \qquad\qquad p(t, q) - \ddot q = -r\ \varepsilon\ N_{V(q)}(v)\ ;$$

in fact, if there is no contact the reaction r vanishes by definition ([Mor 11] (2.1)) and otherwise (9) and (10) still apply.

A motion is said to be of **finite type** if it can be decomposed into a finite number of smooth motions defined on intervals $]t_i, t_{i+1}[$ during which either there is no contact or there is a persistent contact, the possibly existing shocks occurring at (some of) the instants t_i. In such a case we may consider the differential inclusion (20), having only to deal with (17) at a finite number of instants. Unfortunately, this will not happen in a general situation: there exist examples showing that no motion of finite type satisfies the dynamics of the problem. This is the case in [Bress 1] example 1°, where every solution has necessarily an accumulation point of (elastic) shocks.

The synthetic formulation of the inelastic shocks problem proposed by Moreau ([Mor 11] §8) not only encompasses motions which are not of finite type but also contains simultaneously the shock condition (17) or (18) and the dynamical condition of the smooth shockless motion (20).

The data are the following: the interval $I = [0, T]$ $(0 < T < +\infty)$, the region L as in (1), the bounded continuous vector field $p\colon I \times E \to E$, the initial point $q_0 \, \varepsilon \, L$ and the initial kinematically admissible right-velocity $u_0 \, \varepsilon \, V(q_0)$. The unknown is an absolutely continuous (even Lipschitz-continuous) function $q: I \to L$ describing the motion of the system (or material point or else) starting from $q(0)=q_0$. Of course, it is equivalent that the velocity v be known. Here, v cannot be simply a Lebesgue integrable function, defined almost everywhere. In fact, for the formulation to be meaningful we must be able to derivate v in some (not too weak) sense. Hence, we require that v have bounded variation. Its right-continuous companion function $u := v^+$ is then an rcbv function that coincides with the right-velocity of q and which also determines q by integration. The precise statement of the problem is:

Problem 1.1. *Find an rcbv function* $u: I \to E$ *such that* u *and the function* q *defined by:*

$$(21) \qquad q(t) = q_0 + \int_0^t u(s)\, ds \quad (t \, \varepsilon \, I) \, ,$$

satisfy the following:

$$(22) \qquad q(0)=q_0 \, ;$$

$$(23) \qquad u(0) = u_0 \, ;$$

$$(24) \qquad q(t) \, \varepsilon \, L \quad (t \, \varepsilon \, I) \, ;$$

$$(25) \qquad u(t) \, \varepsilon \, V(q(t)) \quad (t \, \varepsilon \, I) \, ;$$

$$(26) \qquad p(t, q(t))\, dt - du \; \varepsilon \; N_{V(q(t))}(u(t)) \, ,$$

in the so-called sense of differential measures: there is a (nonunique) positive measure $d\mu$ *over* I *with respect to which the Lebesgue measure* dt *and the Stieltjes measure* du *both possess densities, respectively* $t'_\mu = \dfrac{dt}{d\mu} \, \varepsilon \, L^1(I, d\mu; \mathbb{R}^+)$ *and* $u'_\mu = \dfrac{du}{d\mu} \, \varepsilon \, L^1(I, du; E)$, *such that:*

$$(27) \qquad p(t, q(t))\, t'_\mu(t) - u'_\mu(t) \; \varepsilon \; N_{V(q(t))}(u(t)) \, ,$$

$d\mu$-*almost everywhere in* I.

Notice that (27) does not depend on the "base" measure $d\mu$, because the right-hand side is conical (see [Mor 11] §8). Hence, if in some subinterval J the function u is itself absolutely continuous, then we may take $d\mu = dt$ in restriction to J so that $t'_\mu = 1$ and $u'_\mu = \dot{u} = \ddot{q}$; thus (27) implies that (20) holds almost everywhere in J (with v replaced by u). And if t is an atom of the measure du, with "value" $[u(t)-u^-(t)]\delta_t$, then t is also an atom of $d\mu$, hence $t'_\mu(t)=0$ (Lebesgue measure has no atoms) and $u'_\mu(t) = \beta(u(t)-u^-(t))$, for some

$\beta > 0$; it is easily verified that (27) is then equivalent to (17)-(18), with $v^+ = u$ and $v^- = u^-$. These two cases do not exhaust all the possibilities contained in the formulation: motions which are not of finite type are clearly admissible, since the velocity may be any function of bounded variation.

In §3.2, we prove that there is a solution to Problem 1.1, by employing a discretization technique similar to the one used in Chapter 2. First, a local solution is found in a interval of specified length and *a priori* estimates on this solution are given (Theorem 2.3); then this is shown to ensure the existence of a global solution (Theorem 2.4). The algorithm presented here is in some sense the most elementary or basic one among those that have been proposed by Moreau (e. g., [Mor 12]) so that less mathematical complexity could reasonably be expected. Other considerations, either from the applications' viewpoint or of numerical analysis nature, justify the use of versions of this algorithm that show such advantages as, for instance, not violating the unilateral constraint or having a faster convergence rate on intervals without contact (e. g. by a Runge-Kutta method). In any case, all these algorithms have a remarkable feature [Mor 13]: the curvature effect of the hypersurface S — cf. (2) — is implicitly taken into account, so that there is no need to compute the second derivatives of its defining function f, which are not even supposed to exist.

The question of uniqueness (under classical assumptions on the vector field p) is not treated here. However, the existing literature on this and similar problems points towards a verdict of **non-uniqueness**.

We mention some relevant papers. In [Sch 1-2], Schatzman studies a problem which contains the particular case of the following differential inclusion (in our notation):

$$(28) \qquad\qquad -du \in N_K(q(t)) \, ,$$

where K is a fixed convex set. Under some other assumptions, (28) is the problem of frictionless elastic shocks in the region K. Existence is proved by a regularization procedure of Yosida type. In general it is an ill-posed problem ([Sch 1] p.606): the solution may not depend continuously on the initial position and velocity and moreover it may not be unique. An example is given there of a convex set K on the boundary of which a certain ray is reflected an infinite number of times in the neighbourhood of $q_0=0$ (cf. [Sch 2]; [Tay] p.27-30); for a certain initial velocity u_0, both that ray and the geodesic which is tangent to u_0 are solutions of (28). However, if the boundary of the convex set

is of class C^2 and its gaussian curvature is strictly positive everywhere, then
the solution is unique and of finite type ([Sch 1] Theorem 2, generalized in
[Per]). Another type of nonuniqueness is given by Bressan in [Bress 1],
example 3, and in [Sch 3] there is an example even for $L=[0,+\infty[$. Buttazzo
and Percivale treated this kind of problem by means of Γ-convergence [But-
Per 1-2]. A discussion of uniqueness and regularity is found in [Bress 3,4].

Among the many papers dealing with the related question of unilateral
contact between deformable bodies, hence with infinite number of degrees of
freedom, let us cite [Ame-Pro] on the vibrating string in the presence of an
obstacle (where nonuniqueness is again possible [Cit]) and [Do] on the
longitudinal dynamics of a bar whose end bumps against an obstacle.

En passant, let us note that several second-order differential inclusions
are treated in the literature; for instance, the following hyperbolic one is a
particular case of [Bar] V.1.1 :

$$(29) \qquad -\frac{du}{dt} \varepsilon \ M(u(t)) + A(q(t)) \,,$$

where M is a fixed time-independent nonlinear monotone multifunction and A
is a continuous symmetric linear operator.

Problem 1.1 applies to all situations where it is reasonable to neglect
frictional effects. If this is not the case, the friction usually present at contact
must be incorporated in the analysis. This will be done by adopting the
formulation of Moreau, especially in [Mor 12] §§ 11, 12.

The classical **isotropic Coulomb's law of friction** is assumed to hold. To
each point $q \varepsilon S$ is associated the so-called *friction cone* $C=C(q)$. This is a cone
with vertex at the origin and revolving about the inward normal to L at the
point q with an angle $\alpha(q) \varepsilon \ ]0, \pi/2[$. The inequality $\alpha(q) < \pi/2$ means that
friction is finite in every direction. Introducing the notation

$$(30) \qquad n = n(q) = - \frac{\nabla f(q)}{\| \nabla f(q) \|} \,,$$

we may describe the friction cone as:

$$(31) \quad C=C(q)=\{ v\varepsilon E\!: \ v . n(q) \geq \| v \| \cos \alpha(q) \} = \{ v\varepsilon E : - v . \nabla f(q) \geq c(q) \| v \| \},$$

where $c(q) := \| \nabla f(q) \| \cos \alpha(q)$ is a real function, positive on S and continuous
by hypothesis. Without significant loss of generality, we even assume that $c(q)$
is defined, positive and continuous on the whole space E.

The friction law stipulates the following:

1°)   The reaction force r belongs to the friction cone:

(32) $$r \varepsilon C ;$$

in particular, it is directed towards the interior of the region L, which means that there is no adhesion effect.

2°)   If there is no contact, then there is no reaction:

(33) $$f(q) < 0 \Rightarrow r = 0 .$$

3°)   If $u = \dot{q}^+$ is the right-velocity, then:

(34) $$u . n > 0 \Rightarrow r = 0 ,$$

because in this case the contact necessarily ceases in some time-interval to the right of the considered instant, hence, by (33), $r = 0$ on that interval and the same is true for the right-limit of r.

4°)   If $u . n = 0$, that is if u belongs to the tangent hyperplane $T = T(q)$, then the classical form of the friction law is equivalent to the new condition (see [Mor 12] (11.15) and pp. 75-76):

(35) $$u . n = 0 \Rightarrow - u \varepsilon \operatorname{proj}_T N_C(r) ,$$

the orthogonal projection into T of the outward normal cone to C at r.

This formulation is remarkable namely for abandoning the usual decomposition of the reaction into its normal and tangential components. Notice also that (35) only determines the direction of r: any λr with $\lambda > 0$ will also satisfy (35). This fact is essential to the statement of the problem in terms of differential measures, as presented next.

From Lagrange's equation (19), that is, from $r = du/dt - p(t, q)$ it follows that $r\, dt = du - p(t, q)\, dt$ holds in the sense of measures, in every subinterval where u is smooth. Let us introduce a vector measure which can be called the *reaction measure*:

(36) $$dR = du - p(t, q(t))\, dt .$$

We see that r may be replaced in (32)-(35) by the density

(37) $$r'_\mu = \frac{dR}{d\mu} \; \varepsilon \; L^1(I, d\mu; E) ,$$

where $d\mu$ is any positive measure with respect to which dR is absolutely continuous. The sought-for mathematical formulation is then:

Problem 1.2. *Given $q_0 \, \varepsilon \, L$ and $u_0 \, \varepsilon \, V(q_0)$, find an rcbv function (the right-velocity) $u \colon I \to E$ defining a Lipschitz-continuous motion $q \colon I \to E$ by integration*

$$(38) \qquad q(t) = q_0 + \int_0^t u(s) \, ds \quad (t \, \varepsilon \, I) \, ,$$

in such a way that:

$$(39) \qquad q(0) = q_0 \, ;$$

$$(40) \qquad u(0) = u_0 \, ;$$

$$(41) \qquad q(t) \, \varepsilon \, L \quad (t \, \varepsilon \, I) \, ;$$

$$(42) \qquad u(t) \, \varepsilon \, V(q(t)) \quad (t \, \varepsilon \, I) \, ;$$

and the following implications are true $d\mu$-almost everywhere:

$$(43) \qquad f(q(t)) < 0 \;\Rightarrow\; r'_\mu(t) = 0 \, ;$$

$$(44) \qquad [f(q(t)) = 0, \; u(t) \cdot \nabla f(q(t)) < 0] \;\Rightarrow\; r'_\mu(t) = 0 \, ;$$

$$(45) \qquad [f(q(t)) = 0, \; u(t) \cdot \nabla f(q(t)) = 0] \;\Rightarrow\; -u(t) \, \varepsilon \, \mathrm{proj}_{T(q(t))} \, N_{C(q(t))}(r'_\mu(t)),$$

where $d\mu$ is any positive measure on I such that r'_μ can be defined by (36) and (37).

Notice that (45) implicitly requires that $r'_\mu(t) \, \varepsilon \, C(q(t))$, $d\mu$-almost everywhere. In case of a shock at time t, the measure dR has an atom at t that equals $(u(t) - u^-(t))\delta_t$ and t is also an atom of the positive measure $d\mu$; since its right-hand side is a cone, (45) is equivalent to

$$(46) \qquad -u(t) \, \varepsilon \, \mathrm{proj}_{T(q(t))} \, N_{C(q(t))}(u(t) - u^-(t)) \, ,$$

which in turn is equivalent to the following (see [Mor 12], p.78-79)

$$(47) \qquad u(t) = \mathrm{proj}\,(\,0\,, \; [u^-(t) + C(q(t))] \cap T(q(t)))\,.$$

The algorithm of approximation developed in §3.3 is based on the last condition (see (3.8)). Under the additional assumption that the gradient of f is a Lipschitz-continuous function of q, it is proved that there exists a solution to Problem 1.2 (defined on I). This is technically more difficult than the frictionless case. Let us point out that (45) is replaced by a "user-friendlier" "variational" condition (Lemma 3.13).

Also in this case, we shall not discuss uniqueness or **non-uniqueness** of solutions, when the force p is Lipschitz-continuous with respect to q. It has long been recognized that, when dealing with the dynamics of systems with dry (Coulomb's) friction, the uniqueness of the motion for certain initial values cannot be ensured [Del 2] [Löt 2] [Bress 2]. This question is linked to the uncertainty concerning the contacts that cease [Del 1] and to the occurence of shocks namely the so-called tangential shocks (discontinuities of velocity after smooth motion episodes and without new contacts appearing); here, we refer of course to the more general case of several unilateral constraints $f_i(q) \leq 0$. In [Del 2] are given some examples where the only possible motions are those with shocks; see also [Löt 2] and [Jea-Pra], last section. Stick-slip phenomena appear e. g. in [Jea-Mor] §§ 6, 8.

These difficulties prompted various authors to find conditions under which the system does not experience any shocks or does not lose contact, at least in the beginning of the motion (for instance, [Jea-Pra] [Del 2]). Then, existence and uniqueness can be established [Jea-Pra]: the normal component of the reaction is taken as the unknown and a fixed-point technique is used to solve a quasi-variational inequality. A smooth behaviour of the solution is also implicitly assumed *a priori* in [Löt 2], where an algorithm and a local existence result are given. Let us mention other works in this active research field. In [Löt 1], the emphasis is put on the numerical aspects without studying existence. In [Löt 3] a less direct algorithm (involving the explicit computation of some Lagrange multipliers) is exhibited for the motion of a bidimensional system with unilateral contacts. In [Jea] is considered a plane obstacle and a motion with persistent contact of a system of points, which allows some simplifications in the algorithm. In [Rio 1, 2] we may find an existence result for the unidimensional case (one degree of freedom). In [Mau] the normal component is assumed to be known as a function of time. Algorithms are also presented in [Tau 1, 2].

From the theoretical point of view, Curnier shows in [Cur 1] the same concern for a coherent formulation, in a setting that goes beyond Coulomb's friction, but restricted instead to small displacements. Another line of research arises when Clarke's normal cone is substituted for the usual cone as in the works [Pan 1, 2].

To end this introduction, let us stress that the numerical techniques proposed by Moreau, namely in [Mor 12], and used here, overcome some of the past difficulties in that they deal simultaneously with: 1) the nondifferentiable relations induced by the dry friction and the unilaterality of the constraint (which are physically related features) and 2) smooth motion episodes and shocks and even possibly cluster points of shocks.

3.2. Frictionless inelastic shocks

The region L is defined by (1.1), that is, by the inequality $f(q) \leq 0$, where $f: I \to \mathbb{R}$ is a C^1 function with nonzero gradient. We prove the following simple property:

Lemma 2.1. *The multifunction $q \to V(q)$, the tangent cone to L at q (see (1.5)), is lower semicontinuous in E and has closed convex values with nonempty interior.*

Proof. In the interior of L, the multifunction V is lsc because it is constant, equal to E. Let $f(q_0) \geq 0$ and U be an open set that intersects $V(q_0)$ and hence also the interior of this set. If we take $v \varepsilon U$ such that $v . \nabla f(q_0) < 0$, then, by continuity of the gradient, we may find a neighbourhood W of q_0 ensuring that $v . \nabla f(q) < 0$ for all $q \varepsilon W$, whence $U \cap V(q) \neq \emptyset$. $\square$

To avoid unessential technical difficulties, we shall assume that the external forces satisfy the following assumption.

Assumption 2.2. *The continuous vector field $p: I \times E \to E$ is globally bounded, i. e., there is a constant $M > 0$ such that:*

$$(1) \qquad \| p(t, q) \| \leq M \qquad (t \varepsilon I, \ q \varepsilon E) .$$

Theorem 2.3. (Existence of a local solution) *Let $q_0 \varepsilon L$ and $u_0 \varepsilon V(q_0)$ be the initial data for $t = 0$. By Lemmas 2.1 and 2.4.2, we take $\delta > 0$ such that:*

$$(2) \qquad \operatorname{int}\left(\bigcap_{\| q - q_0 \| \leq \delta} V(q) \right) \neq \emptyset$$

and $T' > 0$ defined by:

$$(3) \qquad T' = \min \left\{ T, \frac{\delta}{M} \right\} , \ M' = \| u_0 \| + 2 \, T M .$$

Then, on the interval $I' := [0, T']$, Problem 1.1 has at least one solution q with *right-velocity u that satisfies:*

(4) $$\| u(t) \| \leq \| u_0 \| + Mt$$

(*cf.* [*Mor* 11] §8) *and so by* (1.21):

(5) $$\| q(t) \| \leq \| q_0 \| + \| u_0 \| \, t + \tfrac{1}{2} M t^2 \; .$$

The existence of a global solution is then easily deduced:

Theorem 2.4. (Global existence theorem) *The frictionless inelastic shocks' Problem* 1.1 *has at least one global solution, that is, a solution defined on the interval* $I = [0, \, T]$.

Proof. We shall prove that T belongs to the set J of all the S in I for which there exists a solution u_S, q_S of (1.21)-(1.27) defined on $[0, \, S]$. In fact, we show that J equals I, because it is a nonempty closed-open subset of I.

By the preceding Theorem, $T' \varepsilon J$ and every solution u_S, q_S can be continued to the right of S (the theorem is applied with initial instant $t = S$ instead of $t = 0$, which is irrelevant), so J is open.

In order to show that J is closed, consider a sequence (S_k) in J converging to S and assume that $S_k < S$ for all k (otherwise, the limit S belongs trivially to J). Then, by (5), all the solutions q_k , defined on $[0, \, S_k]$ and with right-velocities u_k , satisfy the estimate:

$$\| q_k(t) \| \leq \| q_0 \| + \| u_0 \| \, S + \tfrac{1}{2} M S^2 \; ;$$

in particular, the sequence of final values $q^k := q_k(S_k)$ is bounded in E. Let $\overline{q}$ be a sublimit of (q^k) and choose $\delta^* > 0$ and an integer k such that:

(6) $$\mathrm{int}\Big(\bigcap_{\| q - \overline{q} \| \, \leq \, 2\delta^*} V(q) \Big) \neq \emptyset \; ;$$

(7) $$\| q^k - \overline{q} \| \leq \delta^* \; ;$$

(8) $$S \leq S_k + \delta^* [\, \| u_0 \| + M(S + 2T)]^{-1} \; .$$

From (6) and (7) it follows that

$$\mathrm{int}\Big(\bigcap_{\| q - q^k \| \, \leq \, \delta^*} V(q) \Big) \neq \emptyset \; .$$

By Theorem 2.3, there is a solution $\tilde{q}$, with velocity $\tilde{u}$ and with initial values

$\tilde{q}(S_k) = q^k$ and $\tilde{u}(S_k) = u^k := u_k(S_k)$, which is defined in the interval

$$[S_k , \min\{T, S_k + \delta^*(\| u^k \| + 2\, TM)^{-1}\}] .$$

By (4), $\| u^k \| \leq \| u_0 \| + MS_k \leq \| u_0 \| + MS$. Thus, it is clear from (8) that the above interval contains $[S_k, S]$. Hence q_k and u_k can be continued, by using $\tilde{q}$ and $\tilde{u}$, to give a solution to the inelastic shocks problem which is defined in $[0, S]$; i.e., $S \in J$ and J is closed. $\qquad\square$

The **proof of Theorem 2.3** will be done in several steps. We begin by constructing a sequence (u_n) of approximants of the velocity and the corresponding sequence (q_n) of approximants of the motion, all defined in the interval I'. For each positive integer n, let us take $h(=h_n) = T'/n$ and $t_{n,i} = i/h = iT'/n$ $(0 \leq i \leq n)$ and let us introduce two finite sequences $(q_{n,i})$ and $(u_{n,i})$ of elements of E:

$$(9) \qquad\qquad\qquad q_{n,0} = q_0 ;$$

$$(10) \qquad\qquad u_{n,0} = \mathrm{proj}\,(u_0 + hp(t_{n,0}, q_{n,0}), V(q_{n,0})) ;$$

$$(11) \qquad\qquad\qquad q_{n,i+1} = q_{n,i} + h\, u_{n,i} ;$$

$$(12) \qquad\quad u_{n,i+1} = \mathrm{proj}\,(u_{n,i} + hp(t_{n,i+1}, q_{n,i+1}), V(q_{n,i+1})) .$$

Then, we define u_n by:

$$(13) \qquad u_n(t) = u_{n,i} \qquad \text{if } t \in [t_{n,i}, t_{n,i+1}[\text{ with } 0 \leq i \leq n-1 ,$$

and $u_n(T') = u_{n,n}$. We define q_n by integration:

$$(14) \qquad\qquad\qquad q_n(t) = q_0 + \int_0^t u_n(s)\, ds .$$

Several properties follow from the definitions above. Notice that:

$$(15) \qquad\qquad u_n(t_{n,i}) = u_{n,i} , \quad q_n(t_{n,i}) = q_{n,i} .$$

From (12) and (1) we know that:

$$\| u_{n,i+1} \| \leq \| u_{n,i} + hp(t_{n,i+1}, q_{n,i+1}) \| \leq \| u_{n,i} \| + hM ;$$

and by induction we obtain from (10):

$$(16) \qquad \| u_{n,i} \| \leq \| u_0 \| + (i+1)hM \leq M' := \| u_0 \| + 2\, TM .$$

By (2) there can be found a fixed closed ball $\bar{B}(a, r)$ which is contained in every $V(q)$ provided that $\| q - q_0 \| \leq \delta$. Since (13), (14) and (16) imply that

$$(17) \qquad \| q_n(t) - q_0 \| \leq \int_0^t \| u_n(s) \| \, ds \leq M't \leq M'T' \leq \delta \, ,$$

it follows from (15) that:

$$(18) \qquad V(q_{n,i}) \supset \overline{B}(a, r) \qquad (n \varepsilon \mathbb{N}; \; 0 \leq i \leq n) \, .$$

Lemma 2.5. (a) *The total variation of u_n in $[0, t]$ is bounded above according to:*

$$(19) \qquad \mathrm{var}(u_n; 0, t) \leq \frac{1}{2r} \left(\| u_{n,0} - a \| + hM \right)^2 + \frac{M^2}{2r} t^2 + Mt(1 + \tfrac{1}{r} \| u_{n,0} - a \|) \, .$$

(b) *There is a constant $c > 0$ such that:*

$$(20) \qquad \forall n \varepsilon \mathbb{N} : \quad \mathrm{var}(u_n; I) \leq c \, .$$

Proof. Lemma 0.4.4 (equation 0.4(6)) and conditions (12) and (18) give

$$\| u_{n, i+1} - a \| \leq \| u_{n,i} + h \, p(t_{n, i+1}, q_{n, i+1}) - a \| \leq \| u_{n,i} - a \| + hM,$$

hence, by induction:

$$(21) \qquad \| u_{n,i} - a \| \leq \| u_{n,0} - a \| + M t_{n,i} \, .$$

By the same argument, we deduce from (10) that:

$$(22) \qquad \| u_{n,0} - a \| \leq \| u_0 + h \, p(0, q_0) - a \| \leq \| u_0 - a \| + Mh \, .$$

Writing $w = u_{n,0} + h \, p(t_{n,1}, q_{n,1})$ we prove that, for all i,

$$(23) \qquad \sum_{j=1}^{i} \| u_{n,j} - u_{n,j-1} \| \leq \frac{1}{2r} \left(\| w - a \|^2 - \| u_{n,i} - a \|^2 \right)$$
$$+ \frac{M^2}{2r} t_{n,i}^2 + M t_{n,i} (1 + \tfrac{1}{r} \| u_{n,0} - a \|) \, .$$

If $i = 1$, we simply apply Lemma 0.4.3; since $u_{n,1} = \mathrm{proj}(w, V(q_{n,1}))$ and $\overline{B}(a, r) \subset V(q_{n,1})$ (by (18)), we have:

$$\| u_{n,1} - u_{n,0} \| \leq \| u_{n,1} - w \| + \| w - u_{n,0} \|$$
$$\leq \frac{1}{2r} (\| w - a \|^2 - \| u_{n,1} - a \|^2) + Mh \, .$$

Assuming that (23) holds by induction, we remark that, for similar reasons, in the next step:

$$\| u_{n, i+1} - u_{n,i} \| \leq \frac{1}{2r} (\| u_{n,i} + h \, p(t_{n, i+1}, q_{n, i+1}) - a \|^2 - \| u_{n, i+1} - a \|^2)$$
$$+ Mh;$$

whence, by (21):

$$\| u_{n,i+1} - u_{n,i} \| \leq \tfrac{1}{2r} \left(\| u_{n,i} - a \|^2 - \| u_{n,i+1} - a \|^2 \right)$$

$$+ \tfrac{M^2}{2r} \left(h^2 + 2\,h\,t_{n,i} \right) + M h \left(1 + \tfrac{1}{r} \| u_{n,0} - a \| \right).$$

Adding this inequality to (23) we obtain the formula corresponding to $i+1$ (because $t_{n,i}{}^2 + h^2 + 2\,h\,t_{n,i} = (t_{n,i} + h)^2 = t_{n,i+1}{}^2$) and this establishes the result.

(a) Let now $t \varepsilon I$ and consider the integer i for which $t \varepsilon [t_{n,i}, t_{n,i+1}[$. By (23):

$$\mathrm{var}\,(u_n \,; 0, t) = \sum_{j=1}^{i} \| u_{n,j} - u_{n,j-1} \|$$

$$\leq \tfrac{1}{2r} \| w - a \|^2 + \tfrac{M^2}{2r}\, t^2 + M\,t\,(1 + \tfrac{1}{r} \| u_{n,0} - a \|)\;;$$

and, since $\| w - a \| \leq \| u_{n,0} - a \| + hM$, we get (19).

(b) In particular, take $t = T'$ in (19) and take account of (22) and $h \leq T'$. Then it is clear that (20) is satisfied, with e. g.

$$c = \tfrac{1}{2r}\left(\| u_0 - a \| + 2\,M\,T' \right)^2 + \tfrac{M^2}{2r}\, T'^2 + M\,T'[1 + \tfrac{1}{r}\left(\| u_0 - a \| + 2\,M\,T' \right)]. \qquad \square$$

The sequence of Lipschitz-continuous functions (q_n) is equicontinuous, as a consequence of (13), (14) and (16) :

$$(24) \qquad \| q_n(t) - q_n(s) \| = \| \int_s^t u_n(\tau)\,d\tau \| \leq M' | t - s |.$$

Moreover, for every n:

$$(25) \qquad q_n(0) = q_0.$$

Thus, by Ascoli-Arzelà's Theorem, (q_n) is a relatively compact sequence for the uniform convergence topology. We prove that any sublimit of (q_n) is a solution in I' to the frictionless inelastic shocks' problem.

Consider then a function $q: I' \to E$ such that:

$$(26) \qquad \lim_{\mathcal{I}} q_n(t) = q(t)$$

uniformly in I', where $\mathcal{I}$ is some increasing sequence of positive integers. Since (u_n) is uniformly bounded by M' and it is bounded in total variation (Lemma 2.5(b)), it follows by Theorem 0.2.1 that it is relatively compact for pointwise convergence. So, replacing $\mathcal{I}$ by a subsequence if need be, we may assume that:

$$(27) \qquad \forall t \varepsilon I': \lim_{\mathcal{I}} u_n(t) = v(t),$$

where v is a bv function.

We define $u := v^+$, the right-limit of v. By the dominated convergence theorem, (14) and (26)-(27) yield:

$$(28) \qquad q(t) = q_0 + \int_0^t v(s)\, ds = q_0 + \int_0^t u(s)\, ds \ ,$$

i.e., (1.21) and hence (1.22). Postponing the proof of (1.23), we now show that:

Proposition 2.6. *Every $q(t)$ with $t \varepsilon I'$ belongs to L, that is:*

$$(29) \qquad f(q(t)) \le 0 \quad (t \varepsilon I') \ .$$

Proof. Let K be an upper bound of the norm of the gradient of f in the ball $\overline{B}(q_0, \delta)$ and define for every $\rho > 0$:

$$\eta(\rho) := \max \left\{ \| \nabla f(q) - \nabla f(q') \| \ : \ q, q' \varepsilon \overline{B}(q_0, \delta) \ , \ \| q - q' \| \le \rho \right\} \ .$$

Since the gradient of f is uniformly continuous in the considered ball, we have:

$$\lim_{\rho \to 0} \eta(\rho) = 0 \ .$$

The next inequality measures in some sense the violation of the unilateral constraint committed by the approximant q_n. If $t \varepsilon [0, t_{n,i}]$, then:

$$(30) \qquad f(q_n(t)) \le h M' K + t_{n,i}\, M'\eta(h M') \ , \quad h = \frac{T'}{n} \ .$$

If $i = 0$, this is trivial because $q_n(0) = q_0 \varepsilon L$. Assuming that (30) holds by induction, we next consider an arbitrary $t \varepsilon \,]t_{n,i}, t_{n,i+1}]$.

If $f(q_{n,i}) \le 0$, then by (16) and (17):

$$f(q_n(t)) = f(q_{n,i}) + \int_{t_{n,i}}^t \nabla f(q_n(s)) \cdot u_n(s)\, ds \le \int_{t_{n,i}}^t \| \nabla f(q_n(s)) \| \ \| u_n(s) \| \, ds$$

$$\le h K M' \ ,$$

whence (30).

On the other hand, if $f(q_{n,i}) > 0$, then $u_{n,i} \cdot \nabla f(q_{n,i}) \le 0$, by construction. We notice also that if $s \varepsilon [t_{n,i}, t]$ then $\| q_n(s) - q_{n,i} \| = (s - t_{n,i}) \| u_{n,i} \| \le h M'$. Since η is nondecreasing, using the induction hypothesis and (15)-(17) we have:

$$f(q_n(t)) = f(q_{n,i}) + \int_{t_{n,i}}^t \nabla f(q_{n,i}) \cdot u_{n,i}\, ds + \int_{t_{n,i}}^t [\nabla f(q_n(s)) - \nabla f(q_{n,i})] \cdot u_{n,i}\, ds$$

$$\le f(q_{n,i}) + M' \int_{t_{n,i}}^t \eta(\| q_n(s) - q_{n,i} \|)\, ds$$

$$\le [h M' K + t_{n,i}\, M'\eta(h M')] + M' h \eta(h M') = h M' K + t_{n,i+1} M'\eta(h M') ,$$

which proves (30) in $[0, t_{n,i+1}]$.

Taking limits as $n \to \infty$ with $n \varepsilon \mathcal{I}$ in the inequality (30), we obtain (29) because $t_{n,i} \leq T'$, $h = T'/n \to 0$ and therefore $\eta(hM') \to 0$.   □

We have just established (1.24) $q(t) \varepsilon L$, which implies (see (1.6)) that (1.25) $u(t) \varepsilon V(q(t))$ holds for every t in I' with the possible exception of T'; for the right endpoint we adopt the convention:

$$(31) \qquad u(T') = v^+(T') := \text{proj}\,(v^-(T'),\, V(q(T'))) \,.$$

The next result is essential in order to prove (1.26) and (1.27) at continuity points of u and v.

Lemma 2.7. Let $0 \leq s \leq t < T'$ and assume that $z \varepsilon V(y)$ for every y in some neighbourhood of the set $q([s,t])$. Then:

$$(32) \quad z.(v(t) - v(s)) \geq \tfrac{1}{2}(\,\|v(t)\|^2 - \|v(s)\|^2) + \int_s^t (z - v(\tau)).p(\tau, q(\tau))\,d\tau \,.$$

Proof. The uniform convergence in (26) ensures that for n large enough the set $q_n([s,t])$ is contained in the said neighbourhood of $q([s,t])$. In particular, if the nodes of the partition that belong to $]s,t]$ are $t_{n,j} < t_{n,j+1} < \ldots < t_{n,k}$ (j and k vary with n) then $z \varepsilon V(q_{n,i})$ ($j \leq i \leq k$). Since $u_{n,i}$ is the projection of $u_{n,i-1} + h\,p(t_{n,i}, q_{n,i})$ in $V(q_{n,i})$, then, by the well-known property of projections, we may write:

$$(z - u_{n,i}).[\,u_{n,i} - (u_{n,i-1} + h\,p(t_{n,i}, q_{n,i}))] \geq 0 \,.$$

Adding up these inequalities, we have:

$$(33) \qquad z. \sum_{i=j}^k (u_{n,i} - u_{n,i-1}) \geq \sum_{i=j}^k u_{n,i}.(u_{n,i} - u_{n,i-1})$$
$$+ \sum_{i=j}^k h\,(z - u_{n,i}) \cdot p(t_{n,i}, q_{n,i}) \,.$$

The left-hand side equals $z.(u_{n,k} - u_{n,j-1}) = z.(u_n(t) - u_n(s))$, while by using 2.3(20):

$$\sum_{i=j}^k u_{n,i} \cdot (u_{n,i} - u_{n,i-1}) \geq \tfrac{1}{2} \sum_{i=j}^k (\,\|u_{n,i}\|^2 - \|u_{n,i-1}\|^2)$$
$$= \tfrac{1}{2}(\,\|u_n(t)\|^2 - \|u_n(s)\|^2) \,.$$

Hence (33) implies that:

$$z.(u_n(t) - u_n(s)) \geq \tfrac{1}{2}(\,\|u_n(t)\|^2 - \|u_n(s)\|^2) + \sum_{i=j}^k h\,(z - u_{n,i}) \cdot p(t_{n,i}, q_{n,i}) \,.$$

To obtain (32) it now suffices to let $n \to \infty$ with $n \in \mathcal{S}$, use (27) and prove that

$$(34) \qquad \lim_{\mathcal{S}} \sum_{i=j}^{k} h\,(z - u_{n,i}) \cdot p(t_{n,i}, q_{n,i}) = \int_{s}^{t} (z - v(\tau)) \cdot p(\tau, q(\tau))\, d\tau .$$

We begin by remarking that, because $h = t_{n,i+1} - t_{n,i}$, the sum in (34) is nothing but the integral

$$\int_{t_{n,j}}^{t_{n,k+1}} \phi_n(\tau)\, d\tau ,$$

where $\phi_n(\tau) := (z - u_{n,i}) \cdot p(t_{n,i}, q_{n,i}) = (z - u_n(\tau)) \cdot p(t_{n,i}, q_{n,i})$ if $\tau \in [t_{n,i}, t_{n,i+1}[$ with $j \leq i \leq k$.

Let us take a compact neighbourhood W of the set $\{(\tau, q(\tau)): \tau \in [s, t]\}$ in $I \times E$. Given $\epsilon > 0$, by uniform continuity of p in W let us choose a positive number ρ such that $\| p(\tau, y) - p(\tau', y') \| \leq \epsilon$ for all (τ, y) and (τ', y') belonging to W with $\{|\tau - \tau'|, \| y - y' \| \} < \rho$.

Let n_0 be such that $T'/n_0 \leq \min \{\rho, \rho/(2M'), \epsilon\}$ and $\| q(\tau) - q_n(\tau) \| \leq \rho/2$ whenever $n \geq n_0$ and $\tau \in [s, t]$. Defining

$$\psi_n(\tau) := (z - u_n(\tau)) \cdot p(\tau, q(\tau)) ,$$

we have, for those values of n and τ:

$$| \phi_n(\tau) - \psi_n(\tau) | \leq \| z - u_n(\tau) \| \; \| p(t_{n,i}, q_n(t_{n,i})) - p(\tau, q(\tau)) \| \leq (\| z \| + M')\epsilon,$$

because $|t_{n,i} - \tau| \leq h = T'/n \leq T'/n_0 \leq \rho$ and, by (24):

$$\| q_n(t_{n,i}) - q(\tau) \| \leq \| q_n(t_{n,i}) - q_n(\tau) \| + \| q_n(\tau) - q(\tau) \|$$

$$\leq M' |t_{n,i} - \tau| + \rho/2 \leq M'T'/n_0 + \rho/2 \leq \rho .$$

Observe that $| \phi_n(\tau) | \leq (\| z \| + \| u_n(\tau) \|) \, \| p(t_{n,i}, q_{n,i}) \| \leq (\| z \| + M') M$ and similarly:

$$(35) \qquad\qquad | \psi_n(\tau) | \leq (\| z \| + M') M .$$

Since moreover $h \leq T'/n_0 \leq \epsilon$, we obtain for $n \geq n_0$:

$$\left| \int_{t_{n,j}}^{t_{n,k+1}} \phi_n(\tau)\, d\tau - \int_{s}^{t} \psi_n(\tau)\, d\tau \right| \leq \int_{t_{n,j}}^{t} | \phi_n - \psi_n |\, d\tau$$

$$+ \int_{t}^{t_{n,k+1}} | \phi_n |\, d\tau + \int_{s}^{t_{n,j}} | \psi_n |\, d\tau$$

$$\leq (\| z \| + M') \, \{ \epsilon\,(t - t_{n,j}) + M[(t_{n,k+1} - t) + (t_{n,j} - s)] \}$$

$$\leq (\| z \| + M') \{ \epsilon(t - s) + 2M\epsilon \} .$$

As ϵ is arbitrary, it follows that:

$$\lim_{\substack{y}} \int_{t_{n,j}}^{t_{n,k+1}} \phi_n(\tau)\, d\tau = \lim_{\substack{y}} \int_s^t \psi_n(\tau)\, d\tau = \int_s^t (z - v(\tau)) \cdot p(\tau, q(\tau))\, d\tau \, ,$$

by the dominated convergence theorem, and using (27) and (35). Hence we have established (34), which ends the proof. $\qquad\square$

Since the formulation of Problem 1.1 allows a certain freedom in the choice of the "base" measure $d\mu$, we may take simply:

$$d\mu := |du| + dt \, ,$$

a positive measure with respect to which both $|du|$ and the Lebesgue measure dt possess densities. Then the measure:

$$\left| d(\tfrac{1}{2} \|u\|^2) \right| = \left| \tfrac{1}{2}(u^+ + u^-) \cdot du \right| = \tfrac{1}{2} \|u^+ + u^-\| \, |du|$$

is also absolutely continuous with respect to $d\mu$. Writing $t'_\mu = \frac{dt}{d\mu}$ and $u'_\mu = \frac{du}{d\mu}$, Jeffery's theorem (see [Mor-Val 1-3] or Theorem 0.1.1) now gives:

$$(36) \qquad t'_\mu(t) = \lim_{\epsilon\downarrow 0} \frac{dt([t, t+\epsilon])}{d\mu([t, t+\epsilon])} = \lim_{\epsilon\downarrow 0} \frac{\epsilon}{d\mu([t, t+\epsilon])} \; ;$$

$$(37) \qquad u'_\mu(t) = \lim_{\epsilon\downarrow 0} \frac{du([t, t+\epsilon])}{d\mu([t, t+\epsilon])} \; ;$$

$$(38) \qquad \frac{d(\tfrac{1}{2}\|u\|^2)}{d\mu}(t) = \tfrac{1}{2}\,(u(t) + u^-(t)) \cdot u'_\mu(t) = \lim_{\epsilon\downarrow 0} \frac{d(\tfrac{1}{2}\|u\|^2)([t, t+\epsilon])}{d\mu([t, t+\epsilon])} \; ,$$

for every t outside some $d\mu$-null subset N of I'. We also consider the set

$$N' := \{ t\varepsilon\, I' : \; v^+(t) = v^-(t) \neq v(t) \}$$

which is possibly nonempty (nothing has been said yet about continuity of v). Notice that N' is a countable set, because v is a bv function. Since its points are neither atoms of du, nor obviously of dt, then N' is a $d\mu$-null set.

Proposition 2.8. *If* $t\varepsilon\, I' \backslash (N \cup N')$ *and* u *is continuous at* t, *then* (1.27) *holds, that is:*

$$p(t, q(t))\, t'_\mu(t) - u'_\mu(t) \,\varepsilon\, N_{V(q(t))}(u(t)) \, .$$

Proof. Thanks to the hypotheses, we have:

$$(39) \qquad u(t) := v^+(t) = u^-(t) = v^-(t) = v(t) \, .$$

If z is an interior direction in $V(q(t))$, then the lower semicontinuity of $V(.)$

ensures the existence of an open neighbourhood U of $q(t)$, such that $z \varepsilon V(y)$ for all y in U; we take $\epsilon_0 > 0$ such that $q([t, t+\epsilon_0]) \subset U$. By Lemma 2.7, if $0 < \epsilon_n < \epsilon_0$ then:

$$z . (v(t+\epsilon_n) - v(t)) \geq \tfrac{1}{2}(\| v(t+\epsilon_n) \|^2 - \| v(t) \|^2) + \int_t^{t+\epsilon_n} (z - v(\tau)) . p(\tau, q(\tau)) \, d\tau .$$

We choose the sequence $\epsilon_n \downarrow 0$ in such a way that every $t+\epsilon_n$ satisfies (39). Observe that $v = u$ Lebesgue-almost everywhere in the above integral. Thus, the preceding inequality is equivalent to:

$$z . du([t, t+\epsilon_n]) \geq d(\tfrac{1}{2} \| u \|^2)([t, t+\epsilon_n]) + \int_t^{t+\epsilon_n} (z - u(\tau)) . p(\tau, q(\tau)) \, d\tau .$$

Dividing by $d\mu([t, t+\epsilon_n])$ and passing to the limit as $n \to +\infty$, we obtain:

$$z . u'_\mu(t) \geq u(t) . u'_\mu(t) + \lim_n \left[\frac{1}{d\mu([t, t+\epsilon_n])} \int_t^{t+\epsilon_n} (z - u(\tau)) . p(\tau, q(\tau)) \, d\tau \right] ,$$

thanks to (37)-(39); or equivalently:

$$(z - u(t)) . u'_\mu(t) \geq \lim_n \left\{ \left[\frac{\epsilon_n}{d\mu([t, t+\epsilon_n])} \right] \left[\frac{1}{\epsilon_n} \int_t^{t+\epsilon_n} (z - u(\tau)) . p(\tau, q(\tau)) \, d\tau \right] \right\} .$$

By (36) and the formula for the right-derivative of the integral of a right-continuous function, we deduce that:

$$(z - u(t)) . u'_\mu(t) \geq t'_\mu(t) \, (z - u(t)) . p(t, q(t)) .$$

Hence the inequality

$$(z - u(t)) . [\, p(t, q(t)) \, t'_\mu(t) - u'_\mu(t)] \leq 0$$

holds for every z in the interior of $V(q(t))$; by density, it still holds for every z in the whole of $V(q(t))$ and this proves the result. $\square$

We now study the shocks. We write $v^+ = v^+(t_0) = u(t_0)$ and $v^- = v^-(t_0) = u^-(t_0)$. A shock takes place at time t_0 if $v^- \neq v^+$. A necessary (and also sufficient) condition for a shock to happen is that v^- do not belong to $V(q(t_0))$ since:

Lemma 2.9. *If $v^-(t_0) \varepsilon V(q(t_0))$, then $v^+(t_0) = v^-(t_0)$.*

Proof. If z is an interior point of $V(q(t_0))$, then by letting $s \uparrow t_0$ and $t \downarrow t_0$ in (32) we get:

$$z . (v^+ - v^-) \geq \tfrac{1}{2}(\| v^+ \|^2 - \| v^- \|^2) .$$

By density, this inequality still holds true for $z = v^-$, which by assumption belongs to the cone. Hence,

$$0 \geq \tfrac{1}{2}(\| v^+ \|^2 - 2 v^+ . v^- + \| v^- \|^2) = \tfrac{1}{2} \| v^+ - v^- \|^2 ,$$

that is $v^+ = v^-$. $\qquad\qquad\qquad\qquad\qquad\qquad\qquad\qquad\qquad\qquad\qquad$ $\square$

Remark 2.10. Taking $s \equiv t_0 = 0$ and $v^- = u_0$, this reasoning shows that $v^+(0) = u_0$; that is,

$$(1.23) \qquad\qquad\qquad u(0) = u_0 .$$

Proposition 2.11. *For every $t \varepsilon I$ ', the right-velocity $v^+(t)$ is the projection of the left-velocity $v^-(t)$ in $V(q(t))$; i. e.*

$$(40) \qquad\qquad \forall t \varepsilon I ' : \quad u(t) = \mathrm{proj}(u^-(t), V(q(t))) .$$

Remark 2.12. In particular, at shock instants, which are precisely the atoms of du and hence of $d\mu = | du | + dt$, condition (1.27) is satisfied (see §3.1, (1.17)). So, Propositions 2.8 and 2.11 prove that the inclusion (1.27) holds for $d\mu$-almost every t in I', thus ending the proof of the Existence Theorem 2.3. Observe also that the estimate (4), and *a fortiori* (5), can be directly verified for the constructed solution (start from (13) and (16)).

Proof of Proposition 2.11. If $t = 0$, then by Remark 2.10 and by assumption, $u(0) = u_0 \varepsilon V(q_0)$, while by convention, $u^-(0) = u(0)$; thus (40) holds. If $t = T$', we just use definition (31).

Taking into account Lemma 2.9, we only need to consider the case of a $t_0 \varepsilon \mathrm{int}\, I$' with $v^- \notin V(q(t_0))$. Writing $\hat{q} = q(t_0)$ this means that:

$$f(\hat{q}) = 0 , \quad v^- . \nabla f(\hat{q}) > 0 .$$

We must show that

$$(41) \qquad\qquad v^+ = \hat{v} := \mathrm{proj}(v^-, V(\hat{q})) .$$

Let $\epsilon > 0$. There exists w in the half-space $V(\hat{q})$ such that

$$(42) \qquad\qquad B(w, \epsilon/2) \subset \mathrm{int}\, V(\hat{q})$$

and $\| \hat{v} - w \| = \epsilon$. Notice that if u is close enough to v^- and if y is near to $\hat{q}$ and $f(y) \geq 0$, then, as the gradient of f is a continuous function, we still have

$u \cdot \nabla f(y) > 0$; therefore,

$$\mathrm{proj}\,(u,\ V(y)) = u - (u \cdot \nabla f(y))\,\|\,\nabla f(y)\,\|^{-2}\,\nabla f(y)\ ,$$

a **continuous** function of u and y. Recalling that V is lower semicontinuous, we choose $\rho > 0$ such that the following conditions hold simultaneously:

$$(43) \qquad 0 < \rho < \min\{\ \tfrac{1}{6}\,\|\,v^+ - v^-\,\|\,,\ 3\epsilon\}\ ;$$

$$(44) \qquad [\ \|\,u - v^-\,\| < 3\rho,\ f(y) \geq 0,\ \|\,y - \tilde{q}\,\| < 2\rho\,] \Rightarrow \|\,\mathrm{proj}\,(u,\ V(y)) - \tilde{v}\,\| < \epsilon\ ;$$

$$(45) \qquad \forall\,y\varepsilon\,\bar{B}(\tilde{q},\,2\rho):\quad V(y) \supset \bar{B}(w,\epsilon/2)\ .$$

By definition of side limits and by continuity of q there exists $\eta > 0$ such that:

$$(46) \qquad 0 < \eta < \min\,\{\tfrac{1}{9M}\,\|\,v^+ - v^-\,\|\ ,\ \tfrac{\rho}{3M}\}\ ;$$

$$(47) \qquad \forall t\varepsilon\,[t_0 - \eta,\,t_0[\ :\quad \|\,v(t) - v^-\,\| < \rho\ ;$$

$$(48) \qquad \forall t\varepsilon\,]t_0,\,t_0 + \eta]\ :\quad \|\,v(t) - v^+\,\| < \rho\ ;$$

$$(49) \qquad \forall t\varepsilon\,[t_0 - \eta,\,t_0 + \eta]\ :\quad \|\,q(t) - \tilde{q}\,\| < \rho\ .$$

Moreover, since (q_n) converges uniformly to q and (u_n) converges pointwisely to v, there exists an integer

$$(50) \qquad n_0 \geq \frac{T\,'}{\eta}\,,$$

such that for any $n\varepsilon\,\mathscr{I}$ greater than n_0:

$$(51) \qquad \|\,q_n(t) - q(t)\,\| < \rho \quad (t\varepsilon\,[t_0 - \eta,\,t_0 + \eta])\ ;$$

$$(52) \qquad \|\,u_n(t_0 + \eta) - v(t_0 + \eta)\,\| < \rho\ ;$$

$$(53) \qquad \|\,u_n(t_0 - \eta) - v(t_0 - \eta)\,\| < \rho\ .$$

Fixing n as above, we deduce from (49) and (51) that

$$(54) \qquad \forall t\varepsilon\,[t_0 - \eta,\,t_0 + \eta]\ :\quad \|\,q_n(t) - \tilde{q}\,\| < 2\rho\ ;$$

whereas (48), (52) and (46) imply:

$$(55) \qquad \|\,u_n(t_0 + \eta) - v^+\,\| < 2\rho < \tfrac{1}{3}\,\|\,v^+ - v^-\,\|$$

and analogously, by (47), (53) and (46):

$$(56) \qquad \|\,u_n(t_0 - \eta) - v^-\,\| < 2\rho < \tfrac{1}{3}\,\|\,v^+ - v^-\,\|\ .$$

In turn, (55) and (56) ensure that:

$$(57) \qquad \|\,u_n(t_0 + \eta) - u_n(t_0 - \eta)\,\| > \tfrac{1}{3}\,\|\,v^+ - v^-\,\|\ .$$

Let $\{j, j+1, \ldots, k\}$ be the set of integers for which $t_{n,i}$ belongs to $]t_0 - \eta, t_0 + \eta]$ (it is a nonempty set, because $h = T'/n \leq T'/n_0 \leq \eta$, by (50)). Then $f(q_{n,i}) = f(q_n(t_{n,i})) \geq 0$ for at least one such instant. In fact, if it were not so, then $V(q_{n,i})$ would equal E and $u_{n,i} = u_{n,i-1} + h\,p(t_{n,i}, q_{n,i})$ for all those i, whence

$$\| u_n(t_0 + \eta) - u_n(t_0 - \eta) \| = \| u_{n,k} - u_{n,j-1} \| \leq \sum_{i=j}^{k} \| u_{n,i} - u_{n,i-1} \|$$

$$\leq (k - j + 1) h M \leq 3\eta M < \tfrac{1}{3} \| v^+ - v^- \|$$

by (46), thus contradicting (57).

If t_{n,i_0} is the first of the $t_{n,i}$ $(j \leq i \leq k)$ for which either there is contact or the unilateral constraint is violated, that is $f(q_{n,i_0}) \geq 0$, then we show that:

$$(58) \qquad \| u_{n,i_0} - \tilde{v} \| < \epsilon .$$

We have $u_{n,i_0} = \mathrm{proj}(u, V(q_{n,i_0}))$, with $u := u_{n,i_0-1} + h\,p(t_{n,i_0}, q_{n,i_0})$. By the preceding argument and by (56) and (46), u satisfies:

$$\| u - v^- \| \leq \| h\,p(t_{n,i_0}, q_{n,i_0}) \| + \| u_{n,i_0-1} - u_{n,j-1} \| + \| u_n(t_0 - \eta) - v^- \|$$

$$\leq Mh + M(t_{n,i_0-1} - t_{n,j-1}) + 2\rho \leq 3\eta M + 2\rho \leq 3\rho .$$

Since $f(q_{n,i_0}) \geq 0$, and by (54) $\| q_{n,i} - \tilde{q} \| < 2\rho$, we may apply (44). Then,

$$\| \mathrm{proj}(u, V(q_{n,i_0})) - \tilde{v} \| < \epsilon ,$$

that is, (58). In particular:

$$(59) \qquad \| u_{n,i_0} - w \| < \epsilon + \| \hat{v} - w \| = 2\epsilon .$$

On the other hand, (54) and (45) imply that

$$(60) \qquad V(q_{n,i}) \supset \bar{B}(w, \epsilon/2) \quad (i = j, \ldots, k) .$$

Applying Lemma 2.5(a) to the interval $]t_{n,i_0}, t_0 + \eta]$, with (60) instead of (18), we obtain:

$$(61) \quad \mathrm{var}(u_n; t_{n,i_0}, t_0 + \eta) = \sum_{i=i_0}^{k} \| u_{n,i} - u_{n,i-1} \|$$

$$\leq \tfrac{1}{\epsilon} (\| u_{n,i_0} - w \| + hM)^2 + \tfrac{M^2}{\epsilon} (t_0 + \eta - t_{n,i_0})^2$$

$$+ M(t_0 + \eta - t_{n,i_0}) (1 + \tfrac{2}{\epsilon} \| u_{n,i_0} - w \|)$$

$$\leq 9\epsilon + \tfrac{1}{\epsilon} M^2 (2\eta)^2 + 10\eta M \leq 23\epsilon ,$$

due to (59) and to $hM \leq \eta M \leq \rho/3 \leq \epsilon$, by (43) and (46).

From (58) and (61) we have:

$$\| u_n(t_0 + \eta) - \tilde{v} \| \leq \| u_n(t_0 + \eta) - u_n(t_{n,i_0}) \| + \| u_{n,i_0} - \tilde{v} \| \leq 24\epsilon ,$$

and finally, taking into consideration (55) and (43):

$$\| v^+ - \tilde{v} \| \leq \| v^+ - u_n(t_0 + \eta) \| + 24\,\epsilon < 2\rho + 24\epsilon < 30\epsilon .$$

Since ϵ is arbitrary, we have established (41) $v^+ = \tilde{v}$. $\square$

3.3. Inelastic shocks with friction

This section is devoted to the proof of existence of a **global solution to the Problem 1.2**, under the following assumptions:

Hypotheses 3.1. *The vector field p satisfies assumption 2.2; the angle of the friction cone $C(q)$ is a continuous function $\alpha(q)$ of q, $0 < \alpha(q) < \pi/2$ and $c(q) = \| \nabla f(q) \| \cos \alpha(q)$ is continuous and > 0; and the gradient of the constraint function f is not zero and is Lipschitz-continuous, i. e., there is a positive constant c such that*

$$(1) \qquad \| \nabla f(q) - \nabla f(q') \| \leq c \| q - q' \| \qquad (q,\, q' \varepsilon E) .$$

Let the initial data $q_0 \varepsilon L$ and $u_0 \varepsilon V(q_0)$ be given. We consider a sequence of approximants, defining for every $n \geq 1$:

$$(2) \qquad h = h_n = 2^{-n}\, T ;$$

$$(3) \qquad t_{n,i} = ih = i2^{-n}\, T \quad (i = 0, \ldots, 2^n) ;$$

$$(4) \qquad u_{n,0} = u_0 ;$$

$$(5) \qquad q_{n,0} = q_0 ;$$

$$(6) \qquad q_{n,i} = q_{n,i-1} + h\, u_{n,i-1} \quad (0 < i \leq 2^n) ;$$

$$(7) \qquad u'_{n,i} = u_{n,i-1} + h\, p(t_{n,i},\, q_{n,i}) \quad (0 < i \leq 2^n) ;$$

$$(8) \qquad u_{n,i} = \begin{cases} u'_{n,i} , & \text{if } u'_{n,i} \varepsilon V(q_{n,i}) \\ P(u'_{n,i},\, q_{n,i}) := \mathrm{proj}\,(0, [u'_{n,i} + C(q_{n,i})] \cap T(q_{n,i})), & \text{otherwise;} \end{cases}$$

$$(9) \qquad u_n(t) = \begin{cases} u_{n,i} , & \text{if } t \varepsilon I_{n,i} := [t_{n,i},\, t_{n,i+1}[\text{ with } 0 \leq i < 2^n \\ u_{n,2^n} , & \text{if } t = T = t_{n,2^n} ; \end{cases}$$

$$(10) \qquad q_n(t) = q_0 + \int_0^t u_n(s)\, ds = q_{n,i} + (t - t_{n,i})\, u_{n,i}\,, \quad \text{if } t \varepsilon\, I_{n,i}\,.$$

The first estimates are given in

Lemma 3.2. *For every n and i, we have:*

$$(11) \qquad\qquad \| u_{n,i} \| \leq \| u'_{n,i} \|\,;$$

$$(12) \qquad \| u_{n,i} \| \leq \| u_0 \| + t_{n,i}\, M \leq \tilde{L} := \| u_0 \| + TM\,.$$

Proof. Concerning (11) in the nontrivial case, we remark that the orthogonal projection $w_{n,i}$ of $u'_{n,i}$ in the tangent hyperplane $T(q_{n,i})$ belongs also to the set $u'_{n,i} + C(q_{n,i})$; hence, by definition of $u_{n,i}$:

$$\| u_{n,i} \| \leq \| w_{n,i} \| \leq \| u'_{n,i} \|\,.$$

By induction, (11), (7) and (2.1) immediately imply (12). $\qquad\qquad\square$

Consequently, using definition (10), $\| q_n(t) - q_0 \| \leq T\tilde{L}$; in particular:

$$(13) \qquad\qquad \| q_{n,i} - q_0 \| \leq T\tilde{L}\,.$$

In the closed ball $\bar{B}(q_0, T\tilde{L})$ we can bound the norm of the gradient of f and bound below the coefficient $c(q)$:

$$(14) \qquad c(q) \geq \tilde{c} > 0\,, \text{ if } \| q - q_0 \| \leq T\tilde{L}\,;$$

$$(15) \qquad 0 < l \leq \| \nabla\!f(q) \| \leq G\,, \text{ if } \| q - q_0 \| \leq T\tilde{L}\,.$$

Lemma 3.3. *Denoting $V_{n,i} = V(q_{n,i})$ and $K = G/\tilde{c}$, we have:*

$$(16) \qquad\qquad \| u_{n,i} - u'_{n,i} \| \leq K \operatorname{dist}(u'_{n,i}\,, V_{n,i})\,.$$

Proof. In the nontrivial case, we have $\operatorname{dist}(u'_{n,i}\,, V_{n,i}) = \| u'_{n,i} - w_{n,i} \|$. Note that $u_{n,i} \cdot g_{n,i} = w_{n,i} \cdot g_{n,i} = 0$, where $g_{n,i} := \nabla\!f(q_{n,i})$, and that $u_{n,i} - u'_{n,i}$ belongs to $C(q_{n,i})$. Upon using (1.31) and (13)-(15), we get:

$$\tilde{c}\, \| u_{n,i} - u'_{n,i} \| \leq (u'_{n,i} - u_{n,i}) \cdot g_{n,i} = (u'_{n,i} - w_{n,i}) \cdot g_{n,i}$$

$$\leq G\, \| u'_{n,i} - w_{n,i} \|\,,$$

whence (16). $\qquad\qquad\square$

Next, we bound above the total variation $\operatorname{var}(u_n; t_{n,j}, t_{n,k})$ of u_n in a subinterval $]t_{n,j}, t_{n,k}]$ by considering the various possible cases. We find it convenient to say that if $u_{n,i} = u'_{n,i}$ the corresponding step (at the instant $t_{n,i}$) is of *free type*; otherwise, it is a step of *contact type*.

Case #1. *In the interval $J =]t_{n,j}, t_{n,k}]$ there are only steps of free type.*
Then $\| u_{n,i} - u_{n,i-1} \| = \| h\, p(t_{n,i}, q_{n,i}) \| \leq hM$ and adding up from $j+1$ to k:

$$(17) \qquad \operatorname{var}(u_n; t_{n,j}, t_{n,k}) \leq M(t_{n,k} - t_{n,j}) .$$

Case #2. *In J only the last step is of contact type.*
Then $\| u_{n,k-1} - u_{n,j} \| \leq \operatorname{var}(u_n; t_{n,j}, t_{n,k-1}) \leq M(t_{n,k-1} - t_{n,j})$ (as in (17)) and, by construction, $\| u'_{n,k} - u_{n,k-1} \| \leq hM$; whence

$$\| u'_{n,k} - u_{n,j} \| \leq M(t_{n,k} - t_{n,j}) .$$

Using (16) we have:

$$\| u_{n,k} - u'_{n,k} \| \leq K \operatorname{dist}(u'_{n,k}, V_{n,k}) \leq K[\operatorname{dist}(u_{n,j}, V_{n,k}) + M(t_{n,k} - t_{n,j})]$$

and we easily obtain the formula:

$$(18) \quad \operatorname{var}(u_n; t_{n,j}, t_{n,k}) \leq M(1+K)(t_{n,k} - t_{n,j}) + K \operatorname{dist}(u_{n,j}, V_{n,k}) .$$

Case #3. *In $\bar{J}$ only the endpoints are of contact type.*
Then in (18) we are able to control the distance from $u_{n,j}$ to $V_{n,k}$ (any $k > j$):

$$(19) \qquad \operatorname{dist}(u_{n,j}, V_{n,k}) \leq \frac{c\tilde{L}^2}{l}(t_{n,k} - t_{n,j}) .$$

In fact, in the nontrivial case, $u_{n,j} \cdot g_{n,k} > 0$ and, of course, $u_{n,j} \cdot g_{n,j} = 0$, since $u_{n,j}$ is a velocity after a "contact". Elementary calculation shows that:

$$\operatorname{dist}(u_{n,j}, V_{n,k}) = \frac{(u_{n,j} \cdot g_{n,k})}{\| g_{n,k} \|} \leq \frac{1}{l}[u_{n,j} \cdot (g_{n,k} - g_{n,j})] \leq \frac{\tilde{L}}{l} \| g_{n,k} - g_{n,j} \| ,$$

by (12) and (15). On the other hand (1), (10) and (12) imply that:

$$\| g_{n,k} - g_{n,j} \| \leq c \| q_{n,k} - q_{n,j} \| \leq c\tilde{L}(t_{n,k} - t_{n,j}) .$$

Thus, (19) follows and in this case we may replace (18) by the more precise:

$$(20) \qquad \operatorname{var}(u_n; t_{n,j}, t_{n,k}) \leq R(t_{n,k} - t_{n,j}) ,$$

with $R := M(1+K) + Kc\tilde{L}^2/l$ (no relation to the reaction measure dR).

Lemma 3.4. *In the general case, the following estimate holds:*

$$(21) \quad v(j,k) := \operatorname{var}(u_n; t_{n,j}, t_{n,k}) \leq K \sup_{j < i \leq k} \operatorname{dist}(u_{n,j}, V_{n,i}) + R(t_{n,k} - t_{n,j}) .$$

Proof. If there are no contacts, then by (17)

$$v(j,\,k) \leq M(t_{n,\,k} - t_{n,\,j}) \leq R(t_{n,\,k} - t_{n,\,j}) \ .$$

Otherwise, let $k_1 < \ldots < k_m$ be the subscripts of the steps of contact type that fall inside the considered interval.

If $t_{n,\,j}$ is of free type, then by case #2:

$$v(j,\,k_1) \leq M(1+K)(t_{n,\,k_1} - t_{n,\,j}) + K\,\mathrm{dist}\,(u_{n,\,j},\,V_{n,\,k_1}) \leq R(t_{n,\,k_1} - t_{n,\,j}) + Kd,$$

where

$$(22) \qquad\qquad d := \sup\,\{\,\mathrm{dist}\,(u_{n,\,j},\,V_{n,\,i}):\ i = j+1,\ldots,\,k\}\ .$$

If $t_{n,\,j}$ is of contact type, then by case #3 we have

$$v(j,\,k_1) \leq R\,(t_{n,\,k_1} - t_{n,\,j}) \leq R\,(t_{n,\,k_1} - t_{n,\,j}) + Kd\ ,$$

as above.

For intervals between contact-type instants, we have by (20):

$$v(k_i,\,k_{i+1}) \leq R\,(t_{n,\,k_{i+1}} - t_{n,\,k_i}) \qquad (i = j+1,\ldots,\,k)\ .$$

If $k_m = k$, then $v(k_m,\,k) = 0$; otherwise, by (17):

$$v(k_m,\,k) \leq M\,(t_{n,\,k} - t_{n,\,k_m}) \leq R\,(t_{n,\,k} - t_{n,\,k_m})\ .$$

Even in the worst possible case, summing the preceding inequalities we obtain

$$v(j,\,k) \leq R\,(t_{n,\,k} - t_{n,\,j}) + Kd\ . \qquad\qquad \Box$$

Corollary 3.5. *The sequence (u_n) of the approximants of the velocity is uniformly bounded in variation:*

$$(23) \qquad\qquad \mathrm{var}\,(u_n;0,\,T) \leq K^* := K\,\|\,u_0\,\| + R\,T\ .$$

Proof. We have $\mathrm{dist}\,(u_{n,\,0},\,V_{n,\,i}) \leq \|\,u_0\,\|$, since $u_{n,\,0} = u_0$ and $0 \varepsilon V_{n,\,i}$ for all i. Using (21):

$$\mathrm{var}\,(u_n;0,\,T) = v(0,\,2^n) \leq K\,\|\,u_0\,\| + R\,T\ . \qquad\qquad \Box$$

Corollary 3.6. *If $t_{n,\,j}$ is a step of contact type, then*

$$(24) \qquad\qquad \mathrm{var}\,(u_n;t_{n,\,j},\,t_{n,\,k}) \leq 2\,R(t_{n,\,k} - t_{n,\,j})\ .$$

Proof. It suffices to recall that (19) still holds with $i = j+1, \ldots, k$ substituted for k; hence, in (21) we have $Kd \leq K(c\tilde{L}^2/l)(t_{n,k} - t_{n,j}) \leq R(t_{n,k} - t_{n,j})$. $\qquad \square$

So, (u_n) is uniformly bounded in norm (by (9) and (12): $\| u_n(t) \| \leq \tilde{L}$) and in variation (Corollary 3.5). The sequence of Lipschitz-continuous functions (q_n) is uniformly bounded ($q_n(t) \varepsilon \bar{B}(q_0, T\tilde{L})$) and equicontinuous ($\| q_n(t) - q_n(s) \| \leq \tilde{L} | t - s |$). Hence, as in §3.2, let us consider $\mathcal{I} \subset \mathbb{N}$, a Lipschitz-continuous function $q: I \to E$ and a function of bounded variation $v: I \to E$ such that (q_n) converges uniformly to q and (u_n) converges pointwisely to v, when $n \varepsilon \mathcal{I}$, $n \to +\infty$ (see (2.26), (2.27)).

We define an rcbv function $u: I \to E$ by:

$$
(25) \qquad u(t) := \begin{cases} v^+(t), & \text{if } 0 \leq t < T, \\ v^-(T) = u^-(T) & \text{if } t = T \text{ and } u^-(T) \varepsilon V(q(T)), \\ P(u^-(T), q(T)) & \text{if } t = T \text{ and } u^-(T) \notin V(q(T)). \end{cases}
$$

It follows immediately that (1.38) holds, i. e., $q(t) = q_0 + \int_0^t u(s) \, ds$.

We shall prove that *any pair of functions (q, u) obtained in this manner is a solution to Problem* 1.2.

The initial condition (1.39) $q(0) = q_0$ is trivial; (1.40) will be studied later (Remark 3.11); and to prove (1.41), we begin with the following Lemma.

Lemma 3.7. *For all $n \varepsilon \mathbb{N}$ and $i = 0, \ldots, 2^n$ we have:*

$$
(26) \qquad f(q_{n,i}) \leq G\tilde{L}T2^{-n} + c\tilde{L}^2 T2^{-n} t_{n,i} .
$$

Proof. If $i = 0$, then $f(q_{n,0}) = f(q_0) \leq 0$. Assuming that (26) is true by induction, we consider two cases at the next step. In both cases, for some $0 < \theta < 1$, $q' = q_{n,i} + \theta(q_{n,i+1} - q_{n,i})$ is such that:

$$
f(q_{n,i+1}) = f(q_{n,i}) + \nabla f(q') \cdot (q_{n,i+1} - q_{n,i}) .
$$

a) If $f(q_{n,i}) \leq 0$, then $f(q_{n,i+1}) \leq \| \nabla f(q') \| \, \| q_{n,i+1} - q_{n,i} \| \leq G\tilde{L}h$ and *a fortiori* (26) holds for $i+1$.

b) If $f(q_{n,i}) > 0$, then $(q_{n,i+1} - q_{n,i}) \cdot g_{n,i} = h u_{n,i} \cdot g_{n,i} = 0$ and

$$
\| q' - q_{n,i} \| \leq \| q_{n,i+1} - q_{n,i} \| \leq \tilde{L}h \; (= \tilde{L}T2^{-n}) .
$$

Thus, by (26) and (1), we have:

$$f(q_{n,i+1}) = f(q_{n,i}) + (\nabla f(q') - g_{n,i}) \cdot (q_{n,i+1} - q_{n,i})$$

$$\leq [G\tilde{L}T2^{-n} + c\tilde{L}^2 T2^{-n} t_{n,i}] + (c\tilde{L}h)\tilde{L}T2^{-n}$$

$$= G\tilde{L}T2^{-n} + c\tilde{L}^2 T2^{-n} t_{n,i+1} \, . \qquad \qquad \square$$

Corollary 3.8. *For all $t \varepsilon I$, we have $f(q(t)) \leq 0$; that is:*

(1.41) $$q(t) \varepsilon L \quad (t \varepsilon I) \, .$$

Proof. If t belongs to the set

(27) $$I^* := \{ t_{n,i} : n \geq 1, \, 0 \leq i \leq 2^n \} \, ,$$

then t is one of the nodes $t_{m,j}$ for every sufficiently large m in $\mathcal{I}$ (in view of our choice of the partitions). The preceding Lemma implies that:

$$f(q_m(t)) = f(q_{m,j}) \leq 2^{-m}(G\tilde{L}T + c\tilde{L}^2 T^2) \, .$$

Passing to the limit as $m \to +\infty$, we obtain $f(q(t)) \leq 0$ for all $t \varepsilon I^*$. Since I^* is a dense subset of I and $f(q(\,.\,))$ is a continuous function, the result follows. $\square$

Hence (1.42), i.e., $u(t) \varepsilon V(q(t))$ holds for $0 \leq t < T$. For $t = T$ it is simply a consequence of definition (25).

Next, we prove two Lemmas.

Lemma 3.9. *For every fixed $x \varepsilon E$, the real function $q \to \mathrm{dist}\,(x, V(q))$ is upper semicontinuous.*

Proof. Let $d_0 = \mathrm{dist}\,(x, V(q_0))$ and $\epsilon > 0$. The open ball centered at x with radius $d_0 + \epsilon$ intersects the convex cone $V(q_0)$. Since $q \to V(q)$ is a lower semicontinuous multifunction (Lemma 2.1), we have, for every q in some neighbourhood of q_0 , $B(x, d_0 + \epsilon) \cap V(q) \neq \emptyset$, whence $\mathrm{dist}\,(x, V(q)) < d_0 + \epsilon$. $\square$

Lemma 3.10. *Let $0 < t < T$ and $u^-(t) \varepsilon V(q(t))$.*

(a) *If $\epsilon > 0$, then there exist $\delta > 0$ and $n_0 \varepsilon \mathbb{N}$ such that*

(28) $$\mathrm{var}\,(u_n ; t - \delta, \, t + \delta) \leq (K + R)\,\epsilon \quad (n \geq n_0 , \, n \varepsilon \mathcal{I}) \, .$$

(b) *The functions u and v are both continuous at t: $u(t) = u^-(t) = v(t)$.*

Proof. (a) Let us denote $u^- = u^-(t)$ and $V = V(q(t))$. By the preceding Lemma, we consider a positive number r such that, whenever $\| q - q(t) \| \leq r$,

$$\text{dist}(u^-, V(q)) \leq \text{dist}(u^-, V) + \tfrac{\epsilon}{2} = \tfrac{\epsilon}{2} \ .$$

Next, we take $\delta' \varepsilon\,]0, \epsilon/2[$ such that $s \varepsilon\, I(\delta') := [t - \delta', t + \delta']$ implies that $\| q(s) - q(t) \| \leq r/2$. Moreover, by uniform convergence $q_n \to q$, we choose n_0 in such a way that $\| q_n(s) - q(s) \| \leq r/2$ and so

$$\| q_n(s) - q(t) \| \leq r \quad (n \geq n_0 \,; n \varepsilon\, \mathcal{I} \,; s \varepsilon\, I(\delta')) \ .$$

Picking, if necessary, a smaller δ' and a larger n_0, we may assume that $t - \delta'$ and $t + \delta$ $(0 < \delta \leq \delta')$ are nodes of the considered approximants and that they satisfy:

$$\| v(t - \delta') - u^- \| < \tfrac{\epsilon}{4} \ \text{and} \ \| u_n(t - \delta') - v(t - \delta') \| < \tfrac{\epsilon}{4} \ .$$

If $t - \delta' = t_{n,j}$ and $t + \delta = t_{n,k}$, we see that $\| u_{n,j} - u^- \| < \epsilon/2$ and that, for every i from $j + 1$ to k,

$$\text{dist}(u_{n,j}, V_{n,i}) \leq \| u_{n,j} - u^- \| + \text{dist}(u^-, V(q_{n,i})) \leq \epsilon \ .$$

Applying (21), we finally get :

$$\text{var}(u_n ; t - \delta, t + \delta) \leq \text{var}(u_n ; t - \delta', t + \delta) \leq K\epsilon + R\,(2\delta') \leq (K + R)\epsilon \ .$$

(b) In particular, (28) implies that $\| u_n(t + \delta) - u_n(t - \delta) \| \leq (K + R)\epsilon$, for every large enough $n \varepsilon\, \mathcal{I}$. Thus,

$$(29) \qquad\qquad \| v(t + \delta) - v(t - \delta) \| \leq (K + R)\, \epsilon \ ,$$

by pointwise convergence $u_n \to v$. Letting $\delta \to 0$ and remembering that ϵ is arbitrary, we find that $v^+(t) = v^-(t)$, that is, $u(t) = u^-(t)$. But (28) also implies that $\| v(t + \delta) - v(t) \| \leq (K + R)\, \epsilon$ and we deduce similarly that $v^+(t) = v(t)$. $\qquad\qquad \square$

Remark 3.11. Taking $t = 0$ and $u^- = u_0$, the preceding argument is easily adapted and leads to the result $v^+(0) = v(0)$ $(= \lim u_n(0) = u_0)$. So:

$$(1.40) \qquad\qquad u(0) = u_0 \ .$$

Last but not the least, we have to deal with the implications (1.43), (1.44) and (1.45) – (1.47). Again, we choose the "base" measure:

$$(30) \qquad\qquad d\mu := | du | + dt \ .$$

Notice that $t=0$ is not an atom of $d\mu$ and that if $t=T$ is an atom then it surely satisfies (1.47), by definition (25). Hence, in the remainder of this section we only need to consider interior points of the interval I ($t\varepsilon]0,T[$).

Different situations arise which are treated separately.

A) Interior points of the region L

Introducing the measure

$$(31) \qquad\qquad d\rho := p(t, q(t))\ dt,$$

we ensure, by the classical results on the derivation of measures, that the following equalities hold outside some $d\mu$-null set N :

$$(32) \qquad\qquad u'_\mu(t) = \lim_{\epsilon \to 0^+} \frac{du(I(\epsilon))}{d\mu(I(\epsilon))}\ ;$$

$$(33) \qquad\qquad p(t, q(t))\ t'_\mu(t) = \lim_{\epsilon \to 0^+} \frac{d\rho(I(\epsilon))}{d\mu(I(\epsilon))}\ ,$$

where $I(\epsilon) = [t-\epsilon, t+\epsilon]$. We prove (1.43) or to be more precise:

Proposition 3.12. *If $t \notin N$ and $f(q(t)) < 0$, then the (right-)velocity u is continuous at t and $r'_\mu(t)=0$.*

Proof. The continuity of u at t follows from Lemma 3.10. Let $\epsilon > 0$ and $n_0 \varepsilon \mathbb{N}$ be such that:

$$f(q_n(s)) < 0 \quad (n \geq n_0, n \varepsilon \mathcal{I}, s \varepsilon I(\epsilon))\ .$$

Under these conditions, all the steps in $I(\epsilon)$ are of free type. Hence, for some k and j (depending on n, of course) we have:

$$u_n(t+\epsilon) - u_n(t-\epsilon) = \sum_{i=j}^{k} (u_{n,i} - u_{n,i-1}) = \sum_{i=j}^{k} p(t_{n,i}, q_{n,i})\ (t_{n,i+1} - t_{n,i})$$

$$= \int_{t_{n,j}}^{t_{n,k+1}} p(\hat{\theta}_n(s), q_n(\hat{\theta}_n(s)))\ ds,$$

where $\hat{\theta}_n(s) := \max\{t_{n,i} : 0 \leq i \leq 2^n, t_{n,i} \leq s\}$. The integrands converge uniformly to $p(s, q(s))$, because $\hat{\theta}_n(s) \to s$ and $q_n \to q$ uniformly. The limits of integration converge to $t-\epsilon$ and $t+\epsilon$. Thus,

$$v(t+\epsilon) - v(t-\epsilon) = \int_{t-\epsilon}^{t+\epsilon} p(s, q(s))\ ds\ ,$$

that is, $du(I(\epsilon)) = d\rho(I(\epsilon))$ for sufficiently small ϵ.

Dividing by $d\mu(I(\epsilon))$ and taking limits according to (32) and (33), we obtain:

$$u'_\mu(t) = p(t, q(t)) \, t'_\mu(t) \, ,$$

i.e., $\quad r'_\mu(t) = u'_\mu(t) - p(t, q(t)) \, t'_\mu(t) = 0 \, .$ $\qquad\qquad\qquad\qquad\qquad$ $\Box$

B) The shocks

Let $t \varepsilon \,]0, T[$ be an instant of shock. By Lemma 3.10,

$$(34) \qquad\qquad f(q) = 0 \, , \quad u^- . g > 0 \, ,$$

where $q = q(t)$, $u^- = u^-(t) = v^-(t)$ and $g = \nabla f(q(t))$. Other notations are $u = u(t) = v^+(t)$, $C = C(q(t))$ and $T = T(q(t))$ (clearly, this is not the right endpoint of I!). We only need to prove (1.47), that is:

$$(35) \qquad\qquad u = z := P(u^-, q) = \mathrm{proj}\,(0, (u^- + C) \cap T) \, .$$

Let $\delta > 0$. Observe that $P(u', q')$ is a continuous function of both u' and q' provided that $u' \notin V(q')$ — it has a continuous expression involving $\nabla f(q')$ and $c(q')$, which are continuous by hypothesis; to be precise, for those u':

$$P(u', q') = \max \{\lambda(u', q'), 0\} \, [u' - \mathrm{dist}(u', T') \, \frac{g'}{\|\,g'\,\|}] \, ,$$

with $g' = \nabla f(q')$, $T' = T(q')$, $\mathrm{dist}(u', T') = \dfrac{u' . g'}{\|\,g'\,\|}$, $c' = c(q')$ and

$$\lambda(u', q') = 1 - [\,(\frac{u' . g'}{c'})^2 - \mathrm{dist}(u', C')^2\,]^{1/2} \, [\,\|\,u'\,\|^2 - \mathrm{dist}(u', T')^2\,]^{1/2} \, .$$

Then we choose

$$(36) \qquad\qquad 0 < \epsilon < \|\,u - u^-\,\|$$

in such a way that

$$(37) \qquad [\,\|\,u' - u^-\,\| < 3\epsilon, \|\,q' - q\,\| < \epsilon, \, u' \notin V(q')\,] \;\Rightarrow\; \|\,P(u', q') - z\,\| < \delta.$$

Thanks to the uniform convergence $q_n \to q$ and to the definition of side limits, we now take $\eta > 0$ and $n_0 \varepsilon \mathbb{N}$ so as to have:

$$(38) \qquad \|\,q_n(s) - q\,\| < \epsilon \qquad (n \geq n_0, \, n \varepsilon \mathcal{S}, \, s \varepsilon [t - \eta, t + \eta]) \, ,$$

$$(39) \qquad \|\,v(s) - u^-\,\| < \epsilon \qquad (s \varepsilon [t - \eta, t[) \, ,$$

$$(40) \qquad \|\,v(s) - u\,\| < \epsilon \qquad (s \varepsilon \,]t, t + \eta]) \, .$$

We select $t', t'' \varepsilon \, I^*$ (see (27)) such that $t' < t < t''$ and

$$(41) \qquad\qquad t'' - t' < \min \{\epsilon/M, \eta, \delta\} \, .$$

Then, for every n large enough, there exist $j=j(n)$ and $k=k(n)$ such that $t'=t_{n,j}$ and $t''=t_{n,k}$.

Assume momentarily that, for an infinite number of integers $n \in \mathcal{I}$, the approximant u_n has only steps of free type in the interval $]t',t'']$. Then (17) implies that $\mathrm{var}(u_n;t',t'') \leq M(t''-t') \leq \epsilon$ and in the limit the same happens to the variation of v. In particular, $\|u-u^-\| \leq \epsilon$, contradicting (36).

Hence, starting with some n in $\mathcal{I}$, there are always steps of contact type in $]t',t'']$. Let $t_{n,c}$ be the first one. By (17):

$$(42) \quad \mathrm{var}(u_n;]t',t_{n,c}[)=\mathrm{var}(u_n;t_{n,j},t_{n,c-1}) \leq M(t_{n,c-1}-t_{n,j}) < M(t_{n,c}-t').$$

In particular,

$$(43) \qquad \| u_n^-(t_{n,c})-u_n(t')\| = \| u_{n,c-1}-u_{n,j}\| < M(t_{n,c}-t') < \epsilon.$$

By (39) and (41), $\|v(t')-u^-\| < \epsilon$ so that, due to pointwise convergence $u_n \to v$, we have $\|u_n(t')-u^-\| < \epsilon$, eventually. With (43), this yields $\| u_n^-(t_{n,c})-u^-\| < 2\epsilon$. Hence,

$$u'_{n,c} := u_{n,c-1}+h\,p(t_{n,c},q_{n,c})=u_n^-(t_{n,c})+h\,p(t_{n,c},q_{n,c})$$

with n sufficiently large satisfies:

$$\| u'_{n,c}-u^-\| < 2\epsilon+2^{-n}\,TM < 3\epsilon.$$

Moreover, (38) and (41) imply that $\| q_n(t_{n,c})-q\| < \epsilon$; and $u_n^-(t_{n,c}) \notin V_{n,c}$ because $t_{n,c}$ is of contact type. Thus, $u_{n,c} = P(u'_{n,c},q_{n,c})$ satisfies:

$$(44) \qquad\qquad \| u_{n,c}-z\| < \delta,$$

by (37). By Corollary 3.6 and (41), we have:

$$(45) \qquad \| u_n(t'')-u_{n,c}\| \leq \mathrm{var}(u_n;t_{n,c},t_{n,k}) \leq 2\,R\,(t_{n,k}-t_{n,c}) \leq 2R\delta.$$

From $u_n \to v$, (44) and (45), it now follows that $\| v(t'')-z\| < (1+2R)\delta$, while $\| v(t'')-u\| < \delta$, by (40). Thus, $\| u-z\| \leq 2(1+R)\delta$ and, since δ is arbitrary, we conclude that $u=z$; that is, (35) is proved.

C) A trivial case

We now deal with condition (1.44), assuming that

$$f(q(t)) = 0 \quad \text{and} \quad u(t).\nabla f(q(t)) < 0.$$

Notice that t cannot be a shock instant, because the right-velocity $u(t)$ would then be tangential, as we have just shown. And if $t > 0$, then t cannot be a continuity point of u, since, in that case, from $u(t) \varepsilon V(q(t))$ and $-u^-(t) \varepsilon V(q(t))$ (see (1.6), (1.7)), we would also deduce that $u(t) \varepsilon T(q(t))$.

Hence this case can only occur if $t = 0$, with initial position q_0 on the boundary S and with a nontangential initial velocity u_0. Since $t = 0$ has zero measure for $d\mu$, (44) is obviously satisfied.

D) Contact points with continuous velocity

Let $0 < t < T$ be such that, with the notation of B):

$$(46) \qquad f(q) = 0 , \quad u . g = 0 .$$

By Jeffery's theorem (see Theorem 0.1.1), there is a $d\mu$-null set $N' \subset I$ such that, for every $t \notin N'$, (36), (37) and (38) hold and moreover (as the measures dR, $d\nu := [p(s, q(s)) . u(s)] ds$ and $g . dR$ also have densities with respect to $d\mu$):

$$(47) \qquad r'_\mu(t) = u'_\mu(t) - p(t, q(t)) \, t'_\mu(t) = \lim_{\epsilon \downarrow 0} \frac{dR(J(\epsilon))}{d\mu(J(\epsilon))} ,$$

$$(48) \qquad \lim_{\epsilon \downarrow 0} \frac{d\nu(J(\epsilon))}{d\mu(J(\epsilon))} = \frac{d\nu}{d\mu}(t) = [p(t, q(t)) . u(t)] \, t'_\mu(t) ,$$

$$(49) \qquad \lim_{\epsilon \downarrow 0} \frac{(g . dr)(J(\epsilon))}{d\mu(J(\epsilon))} = g(t) . r'_\mu(t) ,$$

where $J(\epsilon) := [t, t+\epsilon]$.

We shall prove that *if t does not belong to N' then it satisfies* (45); that is, writing $r' = r'_\mu(t)$:

$$(50) \qquad -u \, \varepsilon \, \mathrm{proj}_T \, N_C(r') .$$

The following "variational" characterization will be helpful:

Lemma 3.13. *If $z \varepsilon T$ and $w \varepsilon C$ satisfy the condition*

$$(51) \qquad \forall x \varepsilon \, \mathrm{int} \, C : \quad (w . g)(x . z) \geq (x . g)(w . z),$$

then

$$(52) \qquad z \, \varepsilon \, \mathrm{proj}_T \, N_C(w) .$$

Proof. Let $w = 0$. Since the angle of the friction cone C with the inner normal (1.30) is less than $\pi/2$, we see that the projection of the polar cone

$$N_C(0) = \{ v \varepsilon E : \forall x \varepsilon C, \ v . x \leq 0 \}$$

into T is the hyperplane T itself. Hence (52) is trivial.

Let instead $w \varepsilon C$, $w \neq 0$. Then $w \cdot g < 0$ and, choosing $y := z - \frac{w \cdot z}{w \cdot g}\, g$, it is clear that $z = \mathrm{proj}_T\, y$ (since $y - z$ is orthogonal to T and $z \varepsilon T$). If x is any interior point of C, then:

$$y \cdot (x - w) = z \cdot x - \frac{w \cdot z}{w \cdot g}\,(g \cdot x) \le z \cdot x - z \cdot x = 0 ,$$

using (51). By density, this is true for every $x \varepsilon C$. Hence $y \varepsilon N_C(w)$ and (52) follows. $\square$

Taken together with Lemma 3.13, the next Proposition establishes (50) and ends the proof of the existence of a solution.

Proposition 3.14. *We have:*

(53) $$r' \varepsilon C ;$$

(54) $$\forall x \varepsilon \,\mathrm{int}\, C: \quad (r' \cdot g)(x \cdot u) \le (r' \cdot u)(x \cdot g) .$$

Proof.   a) We prove (53). If $r' = 0$, it is trivial. If $r' \neq 0$ then, for every sufficiently small ϵ, $dR(J(\epsilon)) \neq 0$ and $d\mu(J(\epsilon)) > 0$ (otherwise, (47) and the convention $0/0 = 0$ would give $r' = 0$). Notice also that t has zero measure for $d\mu$ and for dR. We must prove that (see (1.31)):

(55) $$r' \cdot (-g) \ge c(q)\, \| r' \| .$$

Let

(56) $$0 < \delta < c(q)/2 .$$

Since the sequence of gradients $\nabla f(q_n(s))$ converges uniformly to $\nabla f(q(s))$, which is a continuous function of s, then we may take some $\epsilon > 0$ and some n_0 for which $\| \nabla f(q_n(s)) - g \| \le \delta$ for every s in $J(\epsilon)$ and every $n \ge n_0$. In particular, all the $g_{n,i} := \nabla f(q_n(t_{n,i}))$ corresponding to nodes falling inside $]t, t+\epsilon]$, say those with $i = j, \ldots, k$, will satisfy

(57) $$\| g_{n,i} - g \| \le \delta \qquad (n \ge n_0 , \ n \varepsilon \mathcal{G}) .$$

The coefficient $c(q)$ is also a continuous function of q; so, by the same argument, we can ensure that:

(58) $$\| c(q_{n,i}) - c(q) \| \le \delta .$$

By construction, either $r_{n,i} := u_{n,i} - u'_{n,i}$ is zero or it is equal to $P(u'_{n,i}, q_{n,i}) - u'_{n,i} \varepsilon [u'_{n,i} + C(q_{n,i})] - u'_{n,i}$; hence, $r_{n,i}$ belongs to $C(q_{n,i})$. By definition of the friction cone (1.31) and adding up from j to k, we obtain:

$$\sum_{i=j}^{k} r_{n,i} \cdot (-g_{n,i}) \geq \sum_{i=j}^{k} c(q_{n,i}) \, \| r_{n,i} \| \; .$$

By (57) and (58), we deduce that

$$\sum_{i=j}^{k} r_{n,i} \cdot (-g) + \delta \sum_{i=j}^{k} \| r_{n,i} \| \geq [c(q) - \delta] \sum_{i=j}^{k} \| r_{n,i} \| \; ;$$

i. e., with $p_{n,i} := p(t_{n,i}, q_{n,i})$:

(59) $\qquad \displaystyle\sum_{i=j}^{k} (u_{n,i} - u_{n,i-1} - h\,p_{n,i}) \cdot (-g) \geq [c(q) - 2\delta] \sum_{i=j}^{k} \| u_{n,i} - u_{n,i-1} - h p_{n,i} \| \; .$

Let $\theta_n \colon I \to I$ and $p_n \colon I \to E$ be defined by $\theta_n(0) = 0$,

(60) $\qquad \theta_n(s) := t_{n,i} \; , \; \text{if } s \varepsilon \,]t_{n,i-1}, t_{n,i}] \text{ and } 1 \leq i \leq 2^n \; ,$

(61) $\qquad p_n(s) := p(\theta_n(s), q_n(\theta_n(s))) \quad (= p_{n,i} \; , \; \text{if } s \varepsilon \,]t_{n,i-1}, t_{n,i}]) \; .$

Then in (59) we have:

$$\sum_{i=j}^{k} r_{n,i} = \sum_{i=j}^{k} (u_{n,i} - u_{n,i-1} - hp_{n,i}) = u_{n,k} - u_{n,j-1} + \int_{t_{n,j-1}}^{t_{n,k}} p_n(s)\, ds$$

$$= u_n(t+\epsilon) - u_n(t-\epsilon) + \int_{\hat\theta_n(t)}^{\hat\theta_n(t+\epsilon)} p_n(s)\, ds.$$

Let us take limits with respect to n. Since $t_{n,j-1} \to t$, $t_{n,k} \to t+\epsilon$, since $\theta_n(s)$ converges uniformly to s and $p_n(s)$ converges uniformly to $p(s, q(s))$, then we obtain in the limit:

$$u(t+\epsilon) - u^-(t) + \int_{t}^{t+\epsilon} p(s, q(s))\, ds \, ;$$

here, we have used continuity of u at t (see Lemma 3.10 (b)) and also assumed that u is continuous at $t+\epsilon$. Therefore:

(62) $\qquad \displaystyle\lim_{\mathcal{F}} \sum_{i=j}^{k} r_{n,i} = (du - p(s, q(s))\, ds)\, ([t, t+\epsilon]) = dR(J(\epsilon)) \; ,$

hence

(63) $\qquad \displaystyle\liminf_{\mathcal{F}} \sum_{i=j}^{k} \| r_{n,i} \| \geq \lim_{\mathcal{F}} \| \sum_{i=j}^{k} r_{n,i} \| = \| dR(J(\epsilon)) \| \; .$

Thanks to (56), (62) and (63), we infer from (59) that:

$$dR(J(\epsilon)) \cdot (-g) \geq [c(q) - 2\delta] \, \| dR(J(\epsilon)) \| \; ,$$

for every small ϵ such that $t+\epsilon$ is a continuity point of u. Dividing by

$d\mu(J(\epsilon)) > 0$, letting ϵ go to zero and using (47), we get:

$$r'.(-g) \geq [c(q) - 2\delta] \, \|\, r'\,\| \,.$$

As $\delta \to 0$, this produces (55); that is, $r' \epsilon \, C$.

(b) In view of proving (54), we consider $x \epsilon \, \mathrm{int} \, C(q(t))$: $x.(-g) > c(q) \, \| \, x \, \|$. If s is near to t, then we also have $x.(-\nabla\! f(q(s))) > c(q(s)) \, \| \, x \, \|$ and, by uniform convergence, this still holds for $q_n(s)$, starting at some n. In other words, for some n_0 and ϵ_0:

$$(64) \qquad\qquad x \epsilon \, C(q_{n,i}) = C_{n,i} \qquad (0 < \epsilon \leq \epsilon_0 \,;\, n \geq n_0, \, n \epsilon \, \mathcal{I})$$

provided that $t_{n,i} \epsilon \,]t, \, t+\epsilon]$, which happens for, say, $i = j, \, ..., \, k$.

We begin by establishing the **discretized equivalent of** (54), namely:

$$(65) \qquad\qquad (u_{n,i} \cdot x)(g_{n,i} \cdot r_{n,i}) \leq (u_{n,i} \cdot r_{n,i})(g_{n,i} \cdot x) \,.$$

If $r_{n,i} = 0$, then this is trivial. If not, then we are dealing with a step of contact type. Thus $u'_{n,i} \cdot g_{n,i} > 0$ and $u_{n,i} = \mathrm{proj}\,(0, (u'_{n,i} + C_{n,i}) \cap T_{n,i})$ where $T_{n,i} = T(q_{n,i})$. Moreover, $u_{n,i} \cdot g_{n,i} = 0$ and $g_{n,i} \cdot x < 0$, by (64). So, $\lambda := (r_{n,i} \cdot g_{n,i})/(g_{n,i} \cdot x) = -(u'_{n,i} \cdot g_{n,i})/(g_{n,i} \cdot x)$ is positive and it is readily seen that $z := u'_{n,i} + \lambda x$ belongs to $(u'_{n,i} + C_{n,i}) \cap T_{n,i}$ (indeed, $z \cdot g_{n,i} = 0$). Since $u_{n,i}$ is, by definition, the proximal point of 0 in this set, we have $u_{n,i} \cdot (z - u_{n,i}) \geq 0$, whence:

$$(66) \qquad\qquad \lambda\,(u_{n,i} \cdot x) \geq u_{n,i} \cdot (u_{n,i} - u'_{n,i}) = u_{n,i} \cdot r_{n.i} \,.$$

Given the definition of λ, if we multiply (66) by the negative number $g_{n,i} \cdot x$, then we obtain (65).

Now, we consider any $\delta > 0$ and obtain a **δ-approximate version of** (54) (see (84) below). Recalling that $u^-(t) = u(t) \, \epsilon \, V(q(t))$, by Lemma 3.10 (a) we choose $\epsilon_0 > 0$ and $n_0 \epsilon \, \mathcal{I}$ such that

$$(67) \qquad\qquad \mathrm{var}\,(u_n \,;\, t, \, t+\epsilon) \leq \frac{\delta}{2 \, \| \, x \, \|} \,,$$

for every $\epsilon \leq \epsilon_0$ and every $n \geq n_0$ in $\mathcal{I}$. We can also guarantee that, as in (57):

$$(68) \qquad\qquad \| \, g_{n,i} - g \, \| \leq \frac{\delta}{\| \, x \, \|} \,;$$

and thanks to Lemma 3.10 (b) and to pointwise convergence $u_n \to v$:

$$(69) \qquad\qquad \| \, u_n(t) - u \, \| \leq \frac{\delta}{2 \, \| \, x \, \|} \,.$$

From (64) and (68) we get

$$(70) \qquad 0 > g_{n,i} \cdot x \geq g \cdot x - \delta \ .$$

On the other hand,

$$(71) \qquad u_{n,i} \cdot (u_{n,i} - u'_{n,i}) = u_{n,i} \cdot r_{n,i} \leq 0 \ .$$

In fact, in the nontrivial case $r_{n,i} \neq 0$, observing that $-g_{n,i}$ is an interior point of $C(q_{n,i})$, we have by (66):

$$u_{n,i} \cdot r_{n,i} \leq -\lambda \, (u_{n,i} \cdot g_{n,i}) = 0 \ .$$

Thanks to (70) and (71), the right-hand side of (65) is bounded above as follows:

$$(72) \qquad (u_{n,i} \cdot r_{n,i})(g_{n,i} \cdot x) \leq (u_{n,i} \cdot r_{n,i})(g \cdot x - \delta) \ .$$

For $i = j, \dots, k$, from (67) we obtain that $\| u_{n,i} \cdot x - u_n(t) \cdot x \| \leq \delta/2$, while $\| u_n(t) \cdot x - u \cdot x \| \leq \delta/2$ (by (69)); so, $u_{n,i} \cdot x \leq u \cdot x + \delta$. Since $g_{n,i} \cdot r_{n,i} \leq 0$ $(r_{n,i} \varepsilon C_{n,i})$, we have:

$$(73) \qquad (u_{n,i} \cdot x)(g_{n,i} \cdot r_{n,i}) \geq (u \cdot x + \delta)(g_{n,i} \cdot r_{n,i}) \ .$$

Now, we use (65), (72) and (73) and sum from j to k:

$$(74) \qquad (u \cdot x + \delta) \sum_{i=j}^{k} g_{n,i} \cdot r_{n,i} \leq (g \cdot x - \delta) \sum_{i=j}^{k} u_{n,i} \cdot r_{n,i} \ .$$

To obtain an upper bound of the right-hand side of (74) we must give a lower bound of the sum therein, because $g \cdot x - \delta < 0$ $(x \varepsilon C)$. With the use of 2.3(20),

$$(75) \qquad \sum_{i=j}^{k} u_{n,i} \cdot r_{n,i} = \sum_{i=j}^{k} u_{n,i} \cdot (u_{n,i} - u_{n,i-1}) - \sum_{i=j}^{k} (u_{n,i} \cdot p_{n,i}) h$$

$$\geq (\tfrac{1}{2} \| u_n(t+\epsilon) \|^2 - \tfrac{1}{2} \| u_n(t) \|^2) - \int_{t_{n,j}}^{t_{n,k+1}} \hat{p}_n(s) \cdot u_n(s) \, ds,$$

where $\hat{p}_n(s) := p(\hat{\theta}_n(s), q_n(\hat{\theta}_n(s))) = p_{n,i}$ (if $s \varepsilon [t_{n,i}, t_{n,i+1}[)$; $\hat{\theta}_n$ is defined in the proof of Proposition 3.12.

Let $n \to +\infty$ $(n \varepsilon \mathscr{I})$.

The first two terms in the right-hand side of inequality (75) converge to $d(\tfrac{1}{2} \| u \|^2)(J(\epsilon))$, if $t+\epsilon$ is assumed to be a continuity point of u. Regarding the integral, since $\| \hat{p}_n(s) \cdot u_n(s) \| \leq M\hat{L}$, $t_{n,j} \to t$, $t_{n,k} \to t+\epsilon$, $u_n(s) \to v(s)$ and $\hat{p}_n(s) \to p(s, q(s))$, by applying the dominated convergence theorem we obtain in the limit:

$$\int_t^{t+\epsilon} p(s, q(s)) \cdot v(s)\, ds = \int_t^{t+\epsilon} p(s, q(s)) \cdot u(s)\, ds = d\nu(J(\epsilon)) \ .$$

Hence (75) implies

$$(76) \qquad \limsup_{\mathcal{Y}} \Big[(g \cdot x - \delta) \sum_{i=j}^{k} u_{n,i} \cdot r_{n,i} \Big] \leq (g \cdot x - \delta)\, [\, d(\tfrac{1}{2} \| u \|^2) - d\nu](J(\epsilon)) \ .$$

Turning now to the left-hand side of (74), we see that:

$$(77) \qquad \sum_{i=j}^{k} g_{n,i} \cdot r_{n,i} = \sum_{i=j}^{k} g_{n,i} \cdot (u_{n,i} - u_{n,i-1}) - \sum_{i=j}^{k} (g_{n,i} \cdot p_{n,i})\, h \ .$$

Define $g_n(s) := g_{n,i} = \nabla f\!(q_n(t_{n,i})) = \nabla f\!(q_n(\theta_n(s)))$, if $s \in\,]t_{n,i-1}, t_{n,i}]$. Clearly:

$$(78) \qquad\qquad g_n(s) \to g(s) := \nabla f\!(q(s)) \quad \text{uniformly on } I.$$

Therefore:

$$(79) \qquad \sum_{i=j}^{k} g_{n,i} \cdot (u_{n,i} - u_{n,i-1}) = \sum_{i=j}^{k} g_n(t_{n,i}) \cdot du_n(]t_{n,i-1}, t_{n,i}])$$

$$= \int_{]t_{n,j-1}, t_{n,k}]} g_n(s) \cdot du_n(s) = \int_{]t, t+\epsilon]} g_n(s) \cdot du_n(s) \ .$$

By virtue of (67) and (78):

$$(80) \qquad \Big| \int_{]t, t+\epsilon]} [g_n(s) - g(s)] \cdot du_n(s) \Big| \leq \| g_n - g \|_\infty\, \frac{\delta}{2 \| x \|} \to 0 \ ,$$

when $n \to +\infty$, n remaining in $\mathcal{Y}$. But, by Theorem 0.2.2(ii) [0.2(10)] and recalling that g is a continuous function and that both u and v are continuous at t and at $t+\epsilon$, we see that:

$$\lim_{\mathcal{Y}} \int_{]t, t+\epsilon]} g(s) \cdot du_n(s) = \int_{]t, t+\epsilon]} g(s) \cdot dv(s) + g(t) \cdot [v^+(t) - v(t)]$$

$$- g(t+\epsilon) \cdot [v^+(t+\epsilon) - v(t+\epsilon)]$$

$$= \int_{[t, t+\epsilon]} g(s) \cdot du(s) \ .$$

If taken together with (79) and (80), this implies:

$$(81) \qquad \lim_{\mathcal{Y}} \sum_{i=j}^{k} g_{n,i} \cdot (u_{n,i} - u_{n,i-1}) = \int_{[t, t+\epsilon]} g(s) \cdot du(s) \ .$$

The remaining term in (77) may be written in the form:

$$\sum_{i=j}^{k} (g_{n,i} \cdot p_{n,i})\, h = \sum_{i=j}^{k} [g_n(t_{n,i}) \cdot p_n(t_{n,i})](t_{n,i} - t_{n,i-1}) = \int_{t_{n,j-1}}^{t_{n,k}} g_n(s) \cdot p_n(s)\, ds,$$

hence it can be shown to converge to $\int_t^{t+\epsilon} g(s) \cdot p(s, q(s))\, ds$ (use (78) and so on). Combining with (77) and (81), it turns out that:

$$(82) \qquad \lim_{\mathcal{Y}} \sum_{i=j}^{k} g_{n,i} \cdot r_{n,i} = \int_{[t, t+\epsilon]} g(s) \cdot [du(s) - p(s, q(s))\, ds] = (g \cdot dR)(J(\epsilon)) \ .$$

From (74), (76) and (82) it follows that:

$$(83) \qquad (u.x+\delta)\,(g.dR)(J(\epsilon)) \leq (g.x-\delta)\,[d(\tfrac{1}{2}\|u\|^2) - d\nu](J(\epsilon)) \ .$$

Divide (83) by $d\mu(J(\epsilon))$ and let ϵ go to zero, always with u continuous at $t+\epsilon$. Thanks to (2.38) and (47)-(49), this yields

$$(u.x+\delta)\,(g(t).r'_\mu(t)) \ \leq\ (g.x-\delta)\,\{\tfrac{1}{2}[u(t)+u^-(t)].u'_\mu(t) - [p(t,q(t)).u(t)]\,t'_\mu(t)\}$$

$$= (g.x-\delta)\,\{u(t).[u'_\mu(t) - p(t,q(t))\,t'_\mu(t)]\} \ ,$$

that is, with the notation introduced above:

$$(84) \qquad (u.x+\delta)(g.r') \leq (g.x-\delta)(u.r') \ .$$

Letting $\delta \to 0$ in (84), we obtain at last condition (54):

$$(u.x)(g.r') \leq (g.x)(u.r') \ . \qquad\qquad \square$$

Remark 3.15. The assumption 3.1 can be localized: if we only assume that the gradient of f is locally Lipschitz-continuous (e.g. that f is C^2) then the same method provides a local solution to Problem 1.2.

Remark 3.16. When it is reasonable to neglect the forces acting upon the considered system, i.e., to take $p \equiv 0$, then we only need to assume that f be of class C^1. The main difference is in the search for an upper bound of the total variations of the velocity approximants: we may proceed in another manner, closer to the one adopted in §3.2.

Chapter 4

Externally Induced Dissipative Collisions

4.1. Formulation of the problem

In this Chapter, we consider a problem which is related to the so-called standard inelastic shocks, in the general formulation given by Moreau in [Mor 11]. Again, we shall be dealing with a material point or a system (mechanical or otherwise) with finite number of degrees of freedom, which is represented by a point in an Euclidian space E. The scalar product in E is such that the kinetic energy for a motion $q: I \subset \mathbb{R} \to E$ is given by $\frac{1}{2} \| \dot{q} \|^2$. A continuous force field $p: I \times E \to E$, $(t, q) \to p(t, q)$, acts upon the system, which would obey to Lagrange's equation of motion $\ddot{q}(t) = p(t, q(t))$ if it were free. Instead, we suppose that, by means of some external mechanism, the system is subjected to unilateral constraints which can produce collisions, i. e., shocks, and that these are dissipative, purely inelastic.

Unilateral constraints are again geometrically expressed by inequalities of the form

$$(1) \qquad f_\alpha(q) \leq r ,$$

where r is a real number and the f_α form a finite family of $\mathcal{C}^1$ functions, with $\alpha = 1, \dots, \nu$. These constraints can also be viewed as a kind of level functions (*fonctions-seuil*, in french) measuring relevant features of the system.

The constraints are enforced by prescribing for each $t \varepsilon I$ a (possibly empty) subset of $\mathcal{I} := \{1, \dots, \nu\}$ denoted by $A(t)$. We say that the constraints f_α with $\alpha \varepsilon A(t)$ are activated at time t and we call $t \to A(t)$ the **activation multifunction**.

The activation of a single constraint f_α at time t_0, i.e., $A(t_0) = \{\alpha\}$, means that in a right-neighbourhood of t_0 the value of $f_\alpha(q(t))$ should not be greater than its current level $r := f_\alpha(q_0)$, where $q_0 = q(t_0)$. In other words, $q(t)$ should remain in the "lower section"

$$(2) \qquad L_\alpha(r) := \{q \varepsilon E : f_\alpha(q) \leq r\} .$$

A necessary condition is that the right-velocity $u(t_0) := \dot{q}^+(t_0)$ satisfy $\nabla f_\alpha(q_0) \cdot u(t_0) \leq 0$, i. e., that it belong to the set:

(3) $$V_\alpha(q_0) := \{ v \varepsilon E : \ v . \nabla f_\alpha(q) \leq 0 \} ,$$

which is the tangent half-space to $L_\alpha(r)$, if the gradient is not zero. A sufficient condition is that a similar inequality be satisfied for all t in a right-neighbourhood of t_0.

If the left-velocity $u^-(t_0) = \dot q(t_0)$ already belongs to $V_\alpha(q_0)$, then it equals the right-velocity and there is no collision *strictu sensu*. One possibility consists in the velocity being tangential, i.e., it belongs to the tangent hyperplane

(4) $$T_\alpha(q_0) = \{ v \varepsilon E : \ v . \nabla f_\alpha(q) = 0 \} ;$$

we may say that a **smooth contact** occurs. Note that this does not imply a persistent contact with the level surface

(5) $$S_\alpha(r) : = \{ q \varepsilon E : f_\alpha(q) = r \} ;$$

this is clearly shown by the case of a point moving on a straight line, tangent to this surface but locally contained in the admissible region (2). But the velocity may also point inwards: $u^-(t_0) . \nabla f_\alpha(q_0) < 0$, in which case the system is not disturbed and continues its motion away from the level surface and into the region $L_\alpha(r)$. We call this an **ineffectual** (or useless) **activation**.

On the other hand, if $u^-(t_0) \notin V_\alpha(q_0)$ then a **collision** or **shock** necessarily occurs: right-velocity is different from left-velocity. We consider only **purely dissipative** or **inelastic** collisions; according to Moreau's theory (see [Mor 11] or Chapter 3) the right-velocity is then given by

(6) $$u(t_0) = \mathrm{proj}\, (\, u^-(t_0) , \ V_\alpha(q(t_0))\,),$$

the projection or proximal point of the left-velocity in the set of kinematically admissible right-velocities. In this particular case, it is nothing but the orthogonal projection of $u^-(t_0)$ in the tangent hyperplane $T_\alpha(q(t_0))$. Notice that (6) is true even in the absence of collision, because the proximal point is then $u^-(t_0)$ itself. From this formula (6), we get $\| u \| < \| u^- \|$. Hence, the kinetic energy after the collision $\mathcal{E}_c{}^+ = \frac{1}{2}\| u \|^2$ is less than before $\mathcal{E}_c{}^- = \frac{1}{2}\| u^- \|^2$: some of it was dissipated "during" the collision.

Dissipative collisions may even stop the system. Consider $E = \mathbb{R}^2$, $p \equiv 0$ and a material point moving to the right on the x_1-axis and which arrives at the origin at time $t = 0$. Let $A(0) = \{1\}$ and $f_1(x_1 , x_2) = x_1$. Then $u^-(0) = (\lambda , 0)$ with $\lambda > 0$, $V_1(0,0) =]{-\infty} , 0] \times \mathbb{R}$ and so $u(0) = \mathrm{proj}\,((\lambda , 0), V_1(0,0)) = (0,0)$

and the point will remain at rest. This shows that an instantaneous activation may have a lasting effect: here, to take $A(t) \equiv \emptyset$ or $A(t) \equiv \{1\}$ for $t > 0$ gives rise to the same solution. To replace f_1 by $-f_1$ would lead instead to an ineffectual activation: $u(0) = (\lambda, 0) = u^-(0)$. In terms of a mechanical model, these two situations can be easily distinguished if activation is interpreted as the presentation of an **obstacle**

$$(7) \qquad \mathcal{O}_\alpha(r) := \{\, q \varepsilon E \mid r \leq f_\alpha(q) \leq r+\delta \,\} \,,$$

where δ is a positive number we need not prescribe. In the first case, the material point bumps into the "soft wall" $\mathcal{O}_1(0) = [0, \delta] \times \mathbb{R}$ and is stopped; in the latter, it performs a "narrow escape" from a guillotine $[-\delta, 0] \times \mathbb{R}$ falling just behind it. In this manner, if we use a set of different functions f_α activated one at a time, we may play a sort of "dissipative pinball game".

If **more than one constraint** is activated at a general instant t, the formulation of the collision law is somewhat changed. The system is to be kept (tentatively) in the following admissible region

$$(8) \qquad L(A, t, q) := \{\, x \varepsilon E \mid \forall \alpha \varepsilon A(t): \ f_\alpha(x) \leq f_\alpha(q) \,\} = \bigcap_{\alpha \varepsilon A(t)} L_\alpha(f_\alpha(q)) \,,$$

where $q = q(t)$. Thus, the right-velocity $u(t)$ must belong to the convex set

$$(9) \qquad W(A, t, q) := \{\, v \varepsilon E \mid \forall \alpha \varepsilon A(t): \ v . \nabla f_\alpha(q) \leq 0 \,\} = \bigcap_{\alpha \varepsilon A(t)} V_\alpha(q) \,,$$

usually called the *tangent cone* to $L(A, t, q)$ at the point q; it equals E when $A(t)$ is empty. Since the collision is purely dissipative, $u(t)$ is given by the analogue of (6), namely

$$(10) \qquad u(t) = \mathrm{proj}\,(\, u^-(t), \ W(A, t, q(t))\,) \,.$$

By elementary Convex Analysis this is found to be equivalent to

$$(11) \qquad u(t) \varepsilon W(A, t, q(t)) \,,$$

$$(12) \qquad -(u(t) - u^-(t)) \varepsilon N_{W(A, t, q(t))}(u(t)) \,,$$

the outward normal cone to $W(A, t, q(t))$ at the point $u(t)$, i.e., the set of $w \varepsilon E$ such that $w . (v - u(t)) \leq 0$ for every v in $W(A, t, q(t))$.

Let us now formulate the problem in the form of a differential inclusion. The right-velocity u is assumed to be a function of bounded variation and right-continuous (rcbv). We introduce the **reaction measure**

$$(13) \qquad dr := du - p(t, q(t))\, dt,$$

where dt denotes the Lebesgue measure and du the Stieltjes measure of u. We require that the following **differential inclusion**

$$(14) \qquad -dr := p(t, q(t))\, dt - du \; \varepsilon \; N_{W(A,\,t,\,q(t))}(u(t))$$

be satisfied *in the sense of differential measures* (Moreau), which we now explain. Let $d\mu$ be any nonnegative measure with respect to which dt and du are both absolutely continuous, so that $dt = t'_\mu\, d\mu$ and $du = u'_\mu\, d\mu$, where $t'_\mu \varepsilon L^1(I, d\mu; \mathbb{R})$ and $u'_\mu \varepsilon L^1(I, d\mu; E)$ are densities. It is required that

$$(15) \qquad -\frac{dr}{d\mu}(t) = p(t, q(t))\, t'_\mu(t) - u'_\mu(t) \; \varepsilon \; N_{W(A,\,t,\,q(t))}(u(t)),$$

for $d\mu$-almost every t in I. Since the right-hand side is a cone, this only needs to be verified for a single measure $d\mu$ in the said conditions.

This general formulation includes the following special situations:

a) If in some subinterval $J \subset I$ no constraint is activated, then $W(a, t, q(t)) = E$ and the outward normal cone reduces to the zero vector. So (14) gives $-dr = 0$, that is $p(t, q(t))\, dt = du$. Taking $d\mu = dt$ on this interval J, we get

$$p(t, q(t)) = \frac{du}{dt}(t) = \dot{q}(t),$$

i.e., Lagrange's equation.

b) More generally, if the second derivative $\ddot{q}(t)$ exists a.e. and is integrable on some subinterval J, we may take $d\mu = dt$ and (14)-(15) mean that

$$(16) \qquad p(t, q(t)) - \ddot{q}(t) \varepsilon N_{W(A,\,t,\,q(t))}(\dot{q}(t)) \text{ , a.e. on } J.$$

c) If a collision takes place, then du has an atom at the given instant t and so does $d\mu$ under our assumptions. Hence, without loss of generality, we may take $d\mu$ to be equal (in restriction to t) to the Dirac measure δ_t. Then $t'_\mu(t) = 0$ and $u'_\mu(t) = \frac{du}{d\delta_t}(t) = u(t) - u^-(t)$. Thus (15) is equivalent to the collision condition (12).

The following technical hypotheses will be needed in the sequel.

H1 $-$ *For all $\alpha \varepsilon \{1, \dots, \nu\}$ and all $q \varepsilon E$, the gradient $\nabla f_\alpha(q)$ is not zero.*

H2 $-$ *For every $t \varepsilon I := [0, T]$ and every $q \varepsilon E$, the cone $W(A, t, q)$ has nonempty interior.*

H3 $-$ *For every $t \varepsilon I$, a neighbourhood I_t of t can be found such that $A(s) \subset A(t)$ for all s in I_t.*

H4 — *The force field p is continuous and bounded:*

$$(17) \qquad \| p(t, q) \| \leq M \qquad (t \varepsilon I, \ q \varepsilon E) .$$

H5 — *The force field is Lipschitz-continuous with respect to q and there exists a positive constant k such that:*

$$(18) \qquad \| p(t, q) - p(t, q') \| \leq k \| q - q' \| \qquad (t \varepsilon I, \ q \varepsilon E) .$$

Comments. Hypothesis H1 ensures that the level surfaces are really (hyper)surfaces, which are the topological boundaries of the respective sets $L(\alpha, r)$, and that the unit outer normal is given by

$$n_\alpha(q) = \frac{\nabla f_\alpha(q)}{\| \nabla f_\alpha(q) \|} .$$

Hypothesis H2 is a version of the usual cone condition on the regularity of the boundary and is needed for instance to avoid sharp corners. In Mechanics, this type of requirement is sometimes called a "safe load" condition.

Hypothesis H3 tells us that immediately before or after any given instant t we may not activate more constraints than at t. Mathematically, this can be expressed by saying that the activation multifunction A is *upper semicontinuous* from I (with its usual topology) to $\{1, ..., \nu\}$ (with the discrete topology). This is not a stringent restriction on A. Indeed, it allows the treatment of more general motions than the classical "finite-type" ones, where A is piecewise constant; for instance, the accumulation of collision instants is admissible.

Hypotheses H4 and H5 yield existence and uniqueness of solution to Lagrange's equation. They could possibly be weakened, but that is not the essential aim of this study.

Also, we could require these hypotheses to be satisfied only for all q in some neighbourhood of the given initial position q_0; the following main existence theorem would then be replaced by a local existence result.

Theorem 1.1. *Assume H1-H5. Let $q_0 \varepsilon E$ and an admissible initial velocity $u_0 \varepsilon W(A, 0, q_0)$ be given. Then, there exists a Lipschitz-continuous function $q: I \to E$ with $q(0) = q_0$ and such that*

$$(19) \qquad q(t) = q_0 + \int_0^t u(s) \, ds \qquad (t \varepsilon I := [0, T]),$$

where $u: I \to E$ (the right-velocity) is a right-continuous function of bounded variation and satisfies:

$$(20) \qquad u(0) = \dot{q}^+(0) = u_0 \ ;$$

$$(21) \qquad u(t) \, \varepsilon \, W(A, t, q(t)) \qquad (t \varepsilon I) \ ;$$

$$(22) \qquad p(t, q(t)) \, dt - du \, \varepsilon \, N_{W(A, t, q(t))}(u(t)) \ ,$$

in the sense of differential measures explained above.

The proof is given in §4.2 and it relies on Schauder's fixed point theorem and on the existence results for lower semicontinuous sweeping processes of Chapter 2. In §4.3 we discuss the relation of this problem with the problem of inelastic shocks and also the questions of uniqueness and dependence on the data.

4.2. Existence of a solution

Let us take A and f_α ($\alpha = 1, \ldots, \nu$) satisfying the assumptions of Theorem 1.1. To simplify the notation, we shall omit A in expressions such as $W(A, t, q)$.

Lemma 2.1. (a) *The multifunction* $(t, q) \to W(t, q)$ *is lower semicontinuous in* $I \times E$.

(b) *If* $t \to q(t)$ *is a continuous function from I to E, then*

$$(1) \qquad\qquad\qquad t \to W(t, q(t))$$

is a l.s.c. multifunction in I with closed convex values having nonempty interior in E.

Proof. (a) Let U be an open subset of E such that

$$(2) \qquad\qquad\qquad U \cap W(t, q) \neq \emptyset \ .$$

Hypothesis H2 says that the closed convex set $W(t, q)$ has nonempty interior; hence, it is equal to the closure of its interior. Then (2) implies that U also intersects int $W(t, q)$. Let $v \varepsilon U \cap$ int $W(t, q)$. By definition (1.9) and hypothesis H1, this implies that

$$v \cdot \nabla f_\alpha(q) < 0 \qquad (\alpha \, \varepsilon \, A(t)).$$

As all the given functions f_α are $\mathbb{C}^1$, there exists a neighbourhood U_q of q in E, such that for every $y \varepsilon U_q$:

$$(3) \qquad v \cdot \nabla f_\alpha(y) < 0 \qquad (\alpha \varepsilon A(t)) \ .$$

By H3, we can choose I_t , a neighbourhood of t in I such that

$$(4) \qquad A(s) \subset A(t) \qquad (s \varepsilon I_t) \ .$$

Then, condition (3) is satisfied for every s in I_t and every $\alpha \varepsilon A(s) \subset A(t)$. Thus, $v \varepsilon W(s, y)$ and

$$(5) \qquad W(s, y) \cap U \neq \emptyset \qquad (s \varepsilon I_t, \ y \varepsilon U_q) \ ,$$

which proves that $W(.,.)$ is lower semicontinuous in $I \times E$.

(b) It follows immediately from (a), because the composition of a continuous function with a lower semicontinuous multifunction gives a lower semicontinuous multifunction. $\square$

We now begin to construct a **fixed point scheme** inspired by some remarks by Moreau ([Mor 11], end of §8).

Consider the set

$$K := \{ \, q : I \to E \ | \ q(0) = q_0 \, , \ \| q(t) - q(s) \| \leq R \, | t - s | \quad (t, s \varepsilon I) \} \ ,$$

where $R := \| u_0 \| + T M$ is an *a priori* bound for the right-velocity, as shown below. It is clear that K is a **convex** set of absolutely (more precisely, Lipschitz) continuous functions. Since K is by definition equicontinuous, uniformly bounded $(\| q \|_\infty \leq \| q_0 \| + R T)$ and closed, then by Ascoli-Arzelà's theorem K is **compact** in the Banach space of continuous functions $\mathbb{C}(I, E)$ with the uniform convergence norm. The set K may be presented in another way. Let the *primitivation operator* $P : L^\infty(I, dt ; E) \to \mathbb{C}(I, E)$ be defined by

$$(6) \qquad Pu(t) = q_0 + \int_0^t u(s) \, ds \ .$$

We have:

$$(7) \qquad K = \{ Pu \ | \ u \varepsilon L^\infty, \ \| u \|_\infty \leq R \} \ .$$

To every function $q \varepsilon K$ we associate the multifunction

$$(8) \qquad t \to \Gamma_q(t) := W(t, q(t)) - \int_0^t p(s, q(s)) \, ds \ .$$

By virtue of Lemma 2.1(b) and continuity of the integral term, Γ_q is a lower semicontinuous multifunction with closed convex values having nonempty interior. Moreover, $u_0 \varepsilon\, W(0, q_0) = \Gamma_q(0)$. Hence the results of Chapter 2, §4 ensure the existence of a unique rcbv function $v_q : I \to E$ such that

(9) $$v_q(0) = u_0 \ ;$$

(10) $$v_q(t) \varepsilon\, \Gamma_q(t) \quad (t \varepsilon\, I) \ ;$$

(11) $$- dv_q \ \varepsilon\ N_{\Gamma_q(t)}(v_q(t)) \ ,$$

in the sense of differential measures. Let us write

$$v_q = \sigma(q) \ .$$

It is equivalent to say that the function

(12) $$u_q(t) := v_q(t) + \int_0^t p(s, q(s))\, ds \quad (t \varepsilon\, I) \ ,$$

satisfies the following three conditions:

(13) $$u_q(0) = u_0 \ ;$$

(14) $$u_q(t) \varepsilon\, W(t, q(t)) \quad (t \varepsilon\, I) \ ;$$

(15) $$p(t, q(t))\, dt - du_q \ \varepsilon\ N_{W(t, q(t))}(u_q(t)) \ ,$$

again in the sense of Moreau. In fact, (15) is easily shown, by taking $d\mu = |\, dv_q\,| + |\, du_q\,| + dt$ and observing that $du_q = p(t, q(t))\, dt + dv_q$ and $N_\Gamma(v) = N_{\Gamma + x}(v + x)$, for every x in E and every convex set Γ.

We now establish an *a priori* bound on the solution to (13)-(15):

(16) $$\|\, u_q(t)\,\| \ \leq\ R := \|\, u_0\,\| + TM \ .$$

We denote by t'_μ and $u'_{q,\,\mu}$ the densities of dt and du_q with respect to $d\mu$. Then, by (15), we have, for $d\mu$-almost every t:

$$p(t, q(t))\, t'_\mu(t) - u'_{q,\,\mu}(t) \,\varepsilon\, N_{W(t, q(t))}(u_q(t)) \ .$$

Since $W(t, q(t))$ is a convex cone, this gives elementarily:

$$[p(t, q(t))\, t'_\mu(t) - u'_{q,\,\mu}(t)] \cdot u_q(t) = 0 \ ,$$

i.e., $u_q(t) \cdot p(t, q(t))\, t'_\mu(t) = u_q(t) \cdot u'_{q,\,\mu}(t)$, $d\mu$-almost everywhere. Multiplying by $d\mu$, we obtain the following equality of real measures:

(17) $$u_q \cdot p(\,\cdot\,,\, q(\,\cdot\,))\, dt = u_q \cdot du_q \ .$$

Moreover, the inequality

$$\text{(18)} \qquad \qquad \| u_q(t) \| \ \leq \ \| u_q^-(t) \|$$

holds at the continuity points of u_q (trivially). It also holds at any point of discontinuity, because then (15) implies that $u_q(t) = \text{proj}\,(u_q^-(t),\, W(t, q(t)))$ and we know that the projection is nonexpansive and 0 belongs to the set into which the projection is made. Proceeding as in [Mor 11] (8.11)-(8.15), it follows from (17), (18) and H4 that the rcbv function $\phi(t) := \| u_q(t) \|$ satisfies:

$$\phi^- d\phi \ \leq \ \tfrac{1}{2} d(\phi^2) \ = \ \tfrac{1}{2} d(u_q \cdot u_q) \ \leq \ u_q \cdot du_q \ \leq \ \phi M \, dt \ \leq \ \phi^- M \, dt;$$

whence $d\phi \leq M \, dt$, in any subinterval where ϕ^- is never zero. More precisely and without any restriction, we can deduce, as suggested by the last inequality, that $\phi(t) \leq \phi(0) + M t$, i.e.,

$$\| u_q(t) \| \ \leq \ \| u_0 \| + M t$$

and (16) follows.

Fix $R' > R$. It follows from (12), (14) and (16) that $v_q(t)$ belongs to the set

$$\text{(19)} \qquad C_q(t) := [\, W(t, q(t)) \cap \overline{B}(0, R') \,] - \int_0^t p(s, q(s))\, ds \ ,$$

so that, applying the next Lemma, we see that *v_q is also the solution to the sweeping process by C_q* .

Lemma 2.2. *Let $C \subset E$ be a closed convex set and $x \varepsilon C$. If $\| x \| < R'$, then*

$$\text{(20)} \qquad \qquad N_C(x) = N_{C \cap \overline{B}(0, R')}(x) \ .$$

Proof. If $v \varepsilon N_C(x)$, that is, if $v \cdot (y - x) \leq 0$ for all $y \varepsilon C$, then the same is true for every y in the set $C' := C \cap \overline{B}(0, R')$, hence $v \varepsilon N_{C'}(x)$.

Conversely, assume that $v \cdot (z - x) \leq 0$ for every $z \varepsilon C'$ and consider an arbitrary y in C. If $\theta \varepsilon \,]0, 1[$ is sufficiently small, then $z_\theta := x + \theta (y - x)$ still belongs to the convex set C. Moreover $\| z_\theta \| < R'$, because $z_\theta \to x$ when $\theta \to 0$. Thus $z_\theta \varepsilon C'$ and by hypothesis $v \cdot (z_\theta - x) = \theta\, v \cdot (y - x) \leq 0$. Since $\theta > 0$, we have $v \cdot (y - x) \leq 0$ and so $v \varepsilon N_{C'}(x)$. $\qquad\qquad \square$

We prove that *the operator $\sigma : q \to v_q$ is strongly continuous from K into L^∞* .

Consider a sequence of functions $(q_n) \subset K$ that converges uniformly to $q \varepsilon K$. Concerning Hausdorff distances, it will be shown later that

$$(21) \qquad h_n(t) := h\big(W(t, q_n(t)) \cap B', \; W(t, q(t)) \cap B'\big) \to 0 \quad \text{uniformly in } I,$$

where $B' = \overline{B}(0, R')$. Then a simple calculation and H5 yield

$$h\big(C_{q_n}(t), C_q(t)\big) \leq h_n(t) + \left\| \int_0^t p(s, q_n(s))\, ds - \int_0^t p(s, q(s))\, ds \right\|$$

$$\leq h_n(t) + \int_0^t k \, \| q_n(s) - q(s) \| \, ds$$

$$\leq h_n(t) + k\, T \, \| q_n - q \|_\infty \; ;$$

so, the Hausdorff distances $h\big(C_{q_n}(t), C_q(t)\big)$ converge uniformly to zero. Furthermore, since the intersection of l.s.c. multifunctions is still l.s.c. (see [Ber], Ch. VI, §2) both C_{q_n} and C_q are lower semicontinuous multifunctions taking closed convex values with nonempty interior. Then, by Corollary 2.4.14, the solutions to the respective sweeping processes with initial value u_0 satisfy

$$v_{q_n}(t) \to v_q(t) \quad \text{uniformly on } I.$$

The (Picard type) *integral operator* $\mathcal{F} : K \to L^\infty$ defined by

$$(22) \qquad \mathcal{F}(q)(t) := \int_0^t p(s, q(s))\, ds \qquad (q \varepsilon K, t \varepsilon I) ,$$

is also *strongly continuous*. In fact, by H5:

$$\| \mathcal{F}(q)(t) - \mathcal{F}(\tilde{q})(t) \| \leq k \int_0^t \| q(s) - \tilde{q}(s) \| \, ds \leq k\, T \, \| q - \tilde{q} \|_\infty$$

thus

$$(23) \qquad \| \mathcal{F}(q) - \mathcal{F}(\tilde{q}) \|_\infty \leq k\, T \, \| q - \tilde{q} \|_\infty .$$

Therefore, the *"solution operator"* $S : K \to L^\infty$ defined by

$$S(q) := u_q = v_q + \mathcal{F}(q) = (\sigma + \mathcal{F})(q)$$

is *strongly continuous*.

Notice that the primitivation operator $P : L^\infty \to C$ defined by (6) is also strongly (and Lipschitz) continuous:

$$(24) \quad \| Pu - P\tilde{u} \|_\infty = \sup_t \left\| q_0 + \int_0^t u(s)\, ds - \Big(q_0 + \int_0^t \tilde{u}(s)\, ds \Big) \right\| \leq T \, \| u - \tilde{u} \|_\infty .$$

The operator obtained by composition

$$(25) \qquad \Phi(q) := P(S(q)) = P(u_q)$$

is thus strongly continuous from K to $\mathcal{C}(I, E)$ and it takes values in the compact convex set K, thanks to estimate (16) and definition (7) of K. By *Schauder's fixed point theorem*, Φ has at least a fixed point in K, say q. Then

$$q(t) = q_0 + \int_0^t u(s)\, ds,$$

where $u := u_q$ is an rcbv function of t. By (13)-(15), $u(0) = u_0$, $u(t) \varepsilon\, W(t, q(t))$ and

$$p(t, q(t))\, dt - du \,\varepsilon\, N_{W(t, q(t))}(u(t)) \ .$$

This means that we have found a solution to Problem (1.19)-(1.22). Notice that in the process we give estimates for u and q.

To end the proof of Theorem 1.1, we must still establish the *uniform convergence of the Hausdorff distances in* (21). This requires some **geometrical lemmas**.

In (21) we deal with sets of the form

$$(26) \qquad W(t, q) \cap B' = \{\, v\varepsilon E \mid \ \|v\| \le R'; \ \forall \alpha\, \varepsilon\, A(t)\colon\ v.\nabla f_\alpha(q) \le 0 \,\} \ ,$$

which are intersections of a simpler type of set, namely

$$(27) \qquad H(b) := \{\, v\varepsilon E \mid \ \|v\| \le R', \ v.b \le 0 \,\},$$

where $b\varepsilon E$. These are *half-balls* with radius R' orthogonally opposed to the vector b, with the unique exception of $H(0) = \overline{B}(0, R')$.

Lemma 2.3. *If* $\|a\| = \|b\| = 1$, *then the Hausdorff distance between half-balls satisfies:*

$$(28) \qquad h(H(a), H(b)) \le R' \|b - a\| \ .$$

Proof. Let $v\varepsilon H(a)\backslash H(b)$, i.e., $v.a \le 0$, $v.b > 0$. The vector $w := v - (v.b)\,b$ satisfies $w.b = 0$ and $\|w\|^2 = \|v\|^2 - (v.b)^2 \le \|v\|^2 \le R'^2$, because $\|b\| = 1$. Hence, w belongs to $H(b)$ and so

$$\mathrm{dist}(v, H(b)) \le \|v - w\| = v.b = v.a + v.(b - a) \le v.(b - a) \le R' \|b - a\|.$$

We deduce that $e(H(a), H(b)) \leq R' \| b - a \|$. Exchanging b and a, the result follows. $\qquad\qquad\qquad\qquad\qquad\qquad\qquad\qquad\qquad\qquad\qquad\qquad\quad$ $\square$

The next lemma arises by a detailed inspection of the study of the intersection of multifunctions done by Moreau in [Mor 7], §7.

Lemma 2.4. *Let A_i, B_i $(1 \leq i \leq n)$ be convex sets with nonempty interior, all of them contained in the ball $B' = \overline{B}(0, R')$. Let $A := A_1 \cap ... \cap A_n$ and $B := B_1 \cap ... \cap B_n$.*

(a) If A contains a ball with radius r and if for every i

$$(29) \qquad\qquad\qquad e(A_i, B_i) < \tfrac{r}{4} \ ,$$

then we have

$$(30) \qquad\qquad e(A, B) \leq (\tfrac{3}{2} + \tfrac{4R'}{r})^{n-1} \sum_{i=1}^{n} e(A_i, B_i) \, .$$

(b) If A contains a ball with radius r and B contains a ball with radius r' and if for every i

$$(31) \qquad\qquad h(A_i, B_i) < \tfrac{r^*}{4} \ , \quad r^* := \min\{r, r'\}$$

then the following estimate holds:

$$(32) \qquad\qquad h(A, B) \leq (\tfrac{3}{2} + \tfrac{4R'}{r^*})^{n-1} \sum_{i=1}^{n} h(A_i, B_i) \, .$$

Proof. We only need to prove (30), by induction on n.

For $n = 2$, our argument is similar to [Mor 7], Proposition in §7. Let a be the center of a ball of radius r contained in the set $A = A_1 \cap A_2$. In particular, $a \varepsilon A_1$ and $\mathrm{dist}(a, E \backslash A_2) \geq r$. By (29), there is $b_1 \varepsilon B_1$ such that

$$(33) \qquad\qquad\qquad \| a - b_1 \| < \tfrac{r}{4} \, .$$

Since B_2 has nonempty interior, we may use (6) in [Mor 7] (see Proposition 0.4.5) and (29) to obtain

$$(34) \qquad \mathrm{dist}(a, E \backslash B_2) \geq \mathrm{dist}(a, E \backslash A_2) - e(A_2, B_2) > r - \tfrac{r}{4} = \tfrac{3}{4} r \, .$$

On the other hand,

$$(35) \qquad\qquad \mathrm{dist}(a, E \backslash B_2) \leq \| a - b_1 \| + \mathrm{dist}(b_1, E \backslash B_2) \, .$$

Taking together (33)-(35) yields:

$$(36) \qquad\qquad\qquad \mathrm{dist}(b_1, E \backslash B_2) > \tfrac{r}{2} \, .$$

Thus, the ball with radius $r/2$ and with center b_1, belonging to the convex set B_1, is contained in another convex set, namely B_2. So we may apply yet another inequality due to Moreau [Mor 7], (12) (see Proposition 0.4.6). For every x in E, the following inequality holds

$$(37) \qquad \mathrm{dist}\,(x, B_1 \cap B_2) \leq (1 + \tfrac{2}{r} \| x - b_1 \|)[\mathrm{dist}\,(x, B_1) + \mathrm{dist}\,(x, B_2)] \ .$$

When $x \, \varepsilon \, A_1 \cap A_2 \subset B'$, by (33) we have

$$\| x - b_1 \| \leq \| x \| + \| a \| + \| a - b_1 \| < 2R' + \tfrac{r}{4} \ .$$

Therefore, taking the supremum in (37) easily gives:

$$(38) \qquad e(A_1 \cap A_2, B_1 \cap B_2) \leq (\tfrac{3}{2} + \tfrac{4R'}{r})\,[e(A_1, B_1) + e(A_2, B_2)] \ .$$

Induction is straightforward. Write $c := 3/2 + 4R'/r$. Applying repeatedly (38) to the appropriate convex sets, we have, for $n = 3$:

$$
\begin{aligned}
e(A_1 \cap A_2 \cap A_3, B_1 \cap B_2 \cap B_3) \ &\leq \ c[e(A_1 \cap A_2, B_1 \cap B_2) + e(A_3, B_3)] \\
&\leq \ c\{\,c[e(A_1, B_1) + e(A_2, B_2)] + e(A_3, B_3)\} \\
&\leq \ c^2 \sum_{i=1}^{3} e(A_i, B_i)
\end{aligned}
$$

because $c > 1$ implies $c^2 > c$. And so on.   □

Notice also that if $a, b \, \varepsilon \, E \backslash \{0\}$ and $0 < m \leq \min\{\| a \|, \| b \|\}$ then

$$(39) \qquad \left\| \frac{b}{\| b \|} - \frac{a}{\| a \|} \right\| \leq \tfrac{2}{m} \| b - a \| \ .$$

In fact, we have

$$
\begin{aligned}
\left\| \frac{b}{\| b \|} - \frac{a}{\| a \|} \right\| \ &= \ \frac{1}{\| a \| \, \| b \|} \big\| \, \| a \| b - \| b \| a \, \big\| \\
&\leq \ \frac{1}{\| a \| \, \| b \|} [\,\| \, \| a \| b - \| a \| a \, \| + \| (\| a \| - \| b \|) a \| \,] \\
&\leq \ \frac{2}{\| b \|} \| b - a \| \ \leq \ \tfrac{2}{m} \| b - a \| \ .
\end{aligned}
$$

To end this section, we **prove** (21), that is,

$$(40) \qquad\qquad h(\tilde{W}_n(t), \tilde{W}(t)) \to 0 \quad \text{uniformly on } I \,,$$

where $\tilde{W}_n(t) := W(t, q_n(t)) \cap B'$ and $\tilde{W}(t) := W(t, q(t)) \cap B'$. Since $t \to \tilde{W}(t)$ is l.s.c., then, without loss of generality and by the usual argument (see Lemma 2.4.2), we may assume that all the $\tilde{W}(t)$ contain some fixed ball $\bar{B}(x, 2r)$. By assumption, q_n converges uniformly to q and the gradient of every f_α is a continuous function, which is never zero. Hence, a positive number m can be found for which

$$\| \nabla f_\alpha(q_n(t)) \| \geq m, \quad \| \nabla f_\alpha(q(t)) \| \geq m \quad (1 \leq \alpha \leq \nu \,; n \geq 1 \,; t \varepsilon I).$$

As in (27), we define

$$(41) \qquad H_\alpha(q) := \{\, v \varepsilon E \mid \; \| v \| \leq R', \, v \cdot \nabla f_\alpha(q) \leq 0 \,\} = H(\nabla f_\alpha(q)) \,.$$

Using Lemma 2.3 and (39), we get

$$(42) \qquad h(H_\alpha(q_n(t)), H_\alpha(q(t))) \leq R' \| \frac{\nabla f_\alpha(q_n(t))}{\| \nabla f_\alpha(q_n(t)) \|} - \frac{\nabla f_\alpha(q(t))}{\| \nabla f_\alpha(q(t)) \|} \|$$

$$\leq \frac{2R'}{m} \| \nabla f_\alpha(q_n(t)) - \nabla f_\alpha(q(t)) \| \,.$$

But $\nabla f_\alpha(q_n(t)) \to \nabla f_\alpha(q(t))$ uniformly in I. Hence there is N_0 such that

$$(43) \qquad h(H_\alpha(q_n(t)), H_\alpha(q(t))) < \frac{r}{4} \qquad (n \geq N_0) \,.$$

If $A(t) = \emptyset$, then $h(\tilde{W}_n(t), \tilde{W}(t)) = h(B', B') = 0 \to 0$. Otherwise denote by $\nu(t)$ the cardinal of $A(t)$. Thanks to Lemma 2.4 (a), the estimates (42) and (43) then imply:

$$(44) \quad e(\tilde{W}(t), \tilde{W}_n(t)) = e(\bigcap_{\alpha \varepsilon A(t)} H_\alpha(q(t)) , \bigcap_{\alpha \varepsilon A(t)} H_\alpha(q_n(t)))$$

$$\leq (\frac{3}{2} + \frac{4R'}{2r})^{\nu(t)-1} \sum_{\alpha \varepsilon A(t)} e(H_\alpha(q(t)), H_\alpha(q_n(t)))$$

$$\leq \frac{2R'}{m} (\frac{3}{2} + \frac{2R'}{r})^{\nu-1} \sum_{\alpha=1}^{\nu} \| \nabla f_\alpha(q(t)) - \nabla f_\alpha(q_n(t)) \| \,.$$

Hence we may choose $N \geq N_0$ such that for $n \geq N$

$$(45) \qquad e(\tilde{W}(t), \tilde{W}_n(t)) < r \qquad (t \varepsilon I) \,.$$

By equation (6) in [Mor 7] (see Proposition 0.4.5):

$$\text{dist}(x, E \backslash \tilde{W}_n(t)) \geq \text{dist}(x, E \backslash \tilde{W}(t)) - e(\tilde{W}(t), \tilde{W}_n(t)) > 2r - r = r,$$

hence $\overline{B}(x, r) \subset \tilde{W}_n(t)$ for every t in I and all $n \geq N$. Therefore, recalling also (43), we see that Lemma 2.4 (b) applies, with $r^* = \min \{2r, r\} = r$. This yields (we use (42) and (44)):

$$(46) \qquad h(\tilde{W}(t), \tilde{W}_n(t)) \leq \frac{2R'}{m} (\frac{3}{2} + \frac{4R'}{r})^{\nu-1} \sum_{\alpha=1}^{\nu} \| \nabla f_\alpha(q(t)) - \nabla f_\alpha(q_n(t)) \| \,.$$

The sought-for uniform convergence in (21) follows immediately from estimate (46).

4.3. Complements

The problem of induced collisions presented here is closely related to the study of inelastic shocks, in a more general setting than in Chapter 3. Many authors have considered this question (see references in that Chapter or in [Mor 11]) .

In Moreau's formulation, the fixed region where the system is constrained to remain has the form

$$(1) \qquad L := \{\, q \varepsilon E \mid f_\alpha(q) \leq 0 \ \ (\alpha = 1, \dots, \nu) \,\} \,,$$

where the functions $f_\alpha \colon E \to \mathbb{R}$ are of class C^1 and have nonzero gradients. For a motion $q \colon I \to E$ to take place in L, it is necessary and sufficient that, for all t, the right-velocity $u = u(t)$ belongs to the polyhedral convex cone

$$(2) \qquad V(q) := \{\, v \varepsilon E \mid \forall \alpha \varepsilon J(q) \colon \ v . \nabla f_\alpha(q) \leq 0 \,\}$$

where $q = q(t)$ and $J(q)$ is the *set of contacts* at point q:

$$(3) \qquad J(q) := \{\, \alpha \varepsilon \{1, \dots, \nu\} \mid f_\alpha(q) = 0 \,\} \,.$$

The set $V(q)$ is the tangent cone to L at q ; it equals E whenever $J(q)$ is empty, that is, if q is in the interior of L. As usual, the following assumption is made ([Mor 11], (2.6)) :

$$(4) \qquad \forall q \varepsilon L \colon \ \mathrm{int}\ V(q) \neq \emptyset \,.$$

As before, a vector field $p \colon I \times E \to E$ is given, satisfying H4-H5. Then the **standard inelastic shocks'problem** is formulated similarly to Chapter 3 (see [Mor 11],§8, Problem P):

Problem 3.1. *Given $q_0 \varepsilon L$ and $u_0 \varepsilon V(q_0)$, find an rcbv function $u \colon I \to E$ with $u(0) = u_0$ such that, defining $q(t) = q_0 + \int_0^t u(s)\, ds$, we have for every t $u(t) \varepsilon V(q(t))$ $(\Rightarrow q(t) \varepsilon L)$ and the inclusion*

$$(5) \qquad p(t, q(t))\, dt - du \ \varepsilon \ N_{V(q(t))}(u(t)) \,,$$

holds in the sense of differential measures, i.e., of densities with respect to a "base" measure (cf. (2.15)).

For $\nu = 1$, this is Problem 1.1 of Chapter 3, where the existence of a solution was established. For $\nu \geq 2$ it still remains an open problem, but it can be remarked that *a solution q to Problem 3.1 is a solution to Problem 1.1, if we take an appropriate activation multifunction*, namely

(6) $$A(t) = A_q(t) := J(q(t)) = \{\, \alpha \mid 1 \le \alpha \le \nu,\ f_\alpha(q(t)) = 0 \,\},$$

which describes precisely the contacts of solution q. In fact, we have in that case

$$W(A_q, t, q(t)) = \{\, u \mid \forall \alpha \,\varepsilon\, A_q(t) : u \cdot \nabla f_\alpha(q(t)) \le 0 \,\} = V(q(t)).$$

In other words, **a priori** knowledge of the contacts of a prospective solution turns the problem of inelastic shocks into a problem of externally induced collisions. Classical examples seem to forbid such knowledge, which in any case is rendered difficult by the admissibility of phenomena such as accumulation points of shock instants. Nonetheless, this suggests the following

Question: *Can a fixed point scheme be successfully applied to this situation?*

Unfortunately, the answer seems to be *negative*, as the following considerations attempt to explain.

Let us consider, for instance, the set K of Lipschitz-continuous functions in (some subinterval of) the interval I which satisfy $q(0) = q_0$ and $\| q(t) - q(t_0) \| \le R \, | t - s |$ (R to be specified later). We extend definition (6) to functions $q \,\varepsilon\, K$ which may not comply with $q(t) \,\varepsilon\, L$, by writing

(7) $$\tilde{A}_q(t) := \{\, \alpha \mid f_\alpha(q(t)) \ge 0 \,\}.$$

It is clear that $\tilde{A}_q(t) \supset \tilde{A}_q(s)$ for every s in some neighbourhood of t, i.e., that $\tilde{A}_q$ satisfies H3, and that the functions f_α satisfy H1, by assumption. Moreover, (4) easily implies that H2 holds locally:

$$\mathrm{int}\ W(\tilde{A}_q, t, q) \ne \emptyset,$$

for all $t \,\varepsilon\, I' = [0, \delta]$ and all q sufficiently close to the given initial value q_0. Also, $u_0 \,\varepsilon\, V(q_0) = W(\tilde{A}_q, 0, q_0)$ because $\tilde{A}_q(0) = J(q_0)$. Then, for every q fixed in K, by **Theorem 1.1** there exists (at least) one Lipschitz-continuous function $\tilde{q} : I' \to E$ such that $\tilde{q}(0) = q_0$ and

(8) $$\tilde{q}(t) = q_0 + \int_0^t \tilde{u}(s)\, ds,$$

where $\tilde{u} : I' \to E$ is an rcbv function, $\tilde{u}(0) = u_0$, $\tilde{u}(t) \,\varepsilon\, W(\tilde{A}_q(t), t, \tilde{q}(t))$ and

(9) $$p(t, \tilde{q}(t))\, dt - d\tilde{u} \ \varepsilon\ N_{W(\tilde{A}_q, t, \tilde{q}(t))}(\tilde{u}(t)).$$

Therefore, a multifunction S is defined on the set K which associates to every $q \,\varepsilon\, K$ the set $S(q)$ of solutions $\tilde{q}$ of (8)-(9). As in §2, by choosing a suitable Lipschitz constant R, we can enforce that $S(q) \subset K$.

Suppose that q is a fixed point of S, i.e., $q \varepsilon S(q)$. Then, as shown below, q satisfies

$$(10) \qquad\qquad q(t) \varepsilon L \qquad (t \varepsilon I') \, ;$$

thus $\tilde{A}_q(t) = A_q(t) = J(q(t))$ and it follows that q is a solution to Problem 3.1. In order to prove (10), assume by contradiction that $a := f_\alpha(q(t_0)) > 0$ for some t_0 and some α. The set of all $t \varepsilon [0, t_0]$ for which $f_\alpha(q(t)) \geq a$ is closed. We show that its minimum t_1 is 0, thereby contradicting the hypothesis $f_\alpha(q_0) \leq 0$. In fact, if $t_1 > 0$ then there would exist a sufficiently small positive number ϵ such that $f_\alpha(q(t)) > 0$ for all $t \varepsilon [t_1 - \epsilon, t_1]$. So $\alpha \varepsilon \tilde{A}_q(t)$ and $u(t) \varepsilon W(\tilde{A}_q, t, q(t))$ would imply that $u(t) . \nabla f_\alpha(q(t)) \leq 0$ and in particular:

$$f_\alpha(q(t_1 - \epsilon)) = f_\alpha(q(t_1)) - \int_{t_1 - \epsilon}^{t_1} \nabla f_\alpha(q(t)) . u(t) \, dt \geq f_\alpha(q(t_1)) \geq a \, ,$$

contradicting the definition of t_1.

Hence any fixed point of S is a solution to the general inelastic shocks' problem. The question is: **does S have a fixed point?** We would like to apply a fixed point theorem, e.g., the theorem of Tychonoff-Kakutani-Ky Fan. However, *S is not upper semicontinuous on K*, with respect to uniform convergence topology, as shown by the next example.

Example 3.2. Let $E = \mathbb{R}^2$, $\nu = 1$; $f_1(x_1, x_2) = - x_2$, $p(t, q) \equiv 0$; $q_0 = (0, 1)$ and $u_0 = (0, -1)$. Consider the following sequence of functions

$$(11) \qquad\qquad q_n(t) := (t, (1 - t)^2 + \tfrac{1}{n}) \qquad (0 \leq t \leq 2) \, ,$$

which converges uniformly (and even in stronger topologies) to

$$(12) \qquad\qquad q(t) := (t, (1 - t)^2) \, .$$

Since $\tilde{A}_{q_n}(t) \equiv \emptyset$ we have $W(\tilde{A}_{q_n}, t, q) = E$ for all t and $q \varepsilon E$. Hence, (9) implies $- d\tilde{u}_n \varepsilon N_E(\tilde{u}_n(t)) = \{0\}$, and therefore $\tilde{u}_n(t) \equiv \tilde{u}_n(0) = (0, -1)$. In this case, the solution sets $S(q_n)$ are singletons and the solutions are given by

$$(13) \qquad\qquad \tilde{q}_n(t) = S(q_n)(t) = (0, 1 - t) \qquad (0 \leq t \leq 2) \, .$$

These converge in any topology to the function $t \to (0, 1 - t)$, which is not a solution to the problem relative to the limit function q. In fact, $f_1(q(1)) = f_1(1, 0) = 0$ implies $\tilde{A}_q(1) = \{1\}$, so $W(\tilde{A}_q, 1, q) = \{u \mid u . \nabla f_1(q) \leq 0\} = \{u = (u_1, u_2) \mid u_2 \geq 0\}$ for all $q \varepsilon E$, thus making it clear that $\tilde{u}_n(1) \notin W(\tilde{A}_q, 1, q)$. Anyway, it is easily found that

$$(14) \qquad \tilde{q}(t) = S(q)(t) = \begin{cases} (0, 1-t), & \text{if } 0 \leq t \leq 1 \\ (0,0), & \text{if } 1 \leq t \leq 2 \end{cases}.$$

Notice that $\tilde{A}_{q_n}(t)$ and $\tilde{A}_q(t)$ differ only at $t=1$.

This example shows that *a small change in the activation multifunction* $t \to A(t)$ *may cause a remarkable change in the solutions to the problem of induced collisions*, say q_A. This remark seems to discourage considering an even bolder fixed point scheme: find an activation multifunction A such that $\tilde{A}_{q_A} = A$.

If we try to solve Problem 3.1 by the discretization techniques of Chapter 3, then difficulties of similar nature arise, which can be traced back to the fact that *the solution to the general inelastic shocks' problem does not depend continuously on the initial values.*

Example 3.3. Let $E = \mathbb{R}^2$, $f_1(x, y) = -y$ and $f_2(x, y) = x - y$; hence the region where the motion takes place is $L = \{(x, y) : y \geq 0, \ y \geq x\}$. We take $p(t, q) \equiv 0$ and $u(0) = (1, -1)$.

If $q_0(0) = (-1, 1)$, then the solution $q_0(t)$ runs along the diagonal line in the second quadrant until it reaches the origin, where it remains at rest.

If $q_\epsilon(0) = (-1, 1-\epsilon)$ with $0 < \epsilon < 1$, then the solution $q_\epsilon(t)$ runs on a parallel to that diagonal line, arriving at the axis Ox at the point $(1-\epsilon, 0)$, then follows along this axis up to the origin and afterwards it goes along all the diagonal line of the first quadrant.

Hence, if t is large enough, $q_\epsilon(t)$ does not converge to $q_0(t)$ as $\epsilon \to 0$.

Concerning the question of uniqueness of solution for the problem of induced collisions, similarity with inelastic shocks could suggest that nonuniqueness is possible. On the other hand, here we deal with a fixed pre-assigned activation (or contact) multifunction, so that uniqueness should not be discarded right away. The following lines are a unsuccessful attempt to prove a uniqueness result and are meant to call attention upon an interesting technical difficulty.

In order to simplify, we assume that the force field is null $p \equiv 0$ and that the level functions f_α belong to $\mathbb{C}^2$. Let $q = Pu$ and $\tilde{q} = P\tilde{u}$ be two solutions to Problem (1.19)-(1.22). Then $q(0) = \tilde{q}(0) = q_0$, $u(0) = \tilde{u}(0) = u_0$ and

$$- du \,\varepsilon\, N_{W(t,q(t))}(u(t)) \,, \quad - d\tilde{u} \,\varepsilon\, N_{W(t,\tilde{q}(t))}(\tilde{u}(t)) \,.$$

Since uniqueness is a local property, without loss of generality we may take positive numbers k, m, M and r and take $a \,\varepsilon\, E$ such that, for every t and α:

$$\| u(t) \| < M, \; \| \tilde{u}(t) \| < M, \; \| \nabla f_\alpha(q(t)) \| > m, \; \| \nabla f_\alpha(\tilde{q}(t)) \| > m,$$

$$\| q(t) - \tilde{q}(t) \| < \tfrac{rm}{4kM} \,, \; \| \nabla f_\alpha(q(t)) - \nabla f_\alpha(\tilde{q}(t)) \| \leq k \, \| q(t) - \tilde{q}(t) \|$$

and that $C(t) \supset \bar{B}(a, r)$ and $\tilde{C}(t) \supset \bar{B}(a, r)$, where

$$C(t) := W(t, q(t)) \cap \bar{B}(0, M) \,, \quad \tilde{C}(t) := W(t, \tilde{q}(t)) \cap \bar{B}(0, M) \,.$$

These estimates imply that u and $\tilde{u}$ are also the solutions to the sweeping processes by $C(t)$ and $\tilde{C}(t)$. Hence we may apply Theorem 2.4.12, inequality 2.4(31). Thus:

$$(15) \qquad\qquad \| u(t) - \tilde{u}(t) \|^2 \leq \tfrac{2}{r} \mu(t) \| u_0 - a \|^2 =: c\,\mu(t) \,,$$

where $\mu(t) := \max \{ h(C(s), \tilde{C}(s)) \mid 0 \leq s \leq t \}$. Denoting

$$H_\alpha(q) = \{ v \,\varepsilon\, E \mid \| v \| \leq M, \, v \cdot \nabla f_\alpha(q) \leq 0 \},$$

we notice that, as in (2.42)

$$h(H_\alpha(q(t)), H_\alpha(\tilde{q}(t))) \leq \tfrac{2M}{m} \| \nabla f_\alpha(q(t)) - \nabla f_\alpha(\tilde{q}(t)) \| \leq \tfrac{2kM}{m} \| q(t) - \tilde{q}(t) \| < \tfrac{r}{4} \,.$$

Hence, by Lemma 2.4 (b) we get an expression similar to (2.46):

$$(16) \quad h(C(t), \tilde{C}(t)) \leq \tfrac{2M}{m} \, (\tfrac{3}{2} + \tfrac{4M}{r})^{\nu-1} \sum_{\alpha=1}^{\nu} \| \nabla f_\alpha(q(t)) - \nabla f_\alpha(\tilde{q}(t)) \|$$

$$\leq k^* \, \| q(t) - \tilde{q}(t) \| \,,$$

with $k^* = \tfrac{2M}{m} \, (\tfrac{3}{2} + \tfrac{4M}{r})^{\nu-1} \nu \, k$.

Define $\phi(t) := \max \{ \| q(s) - \tilde{q}(s) \| \; : \; 0 \leq s \leq t \}$. Then, by (16), $\mu(t) \leq k^* \phi(t)$ and, writing $\lambda := c\,k^*$, (15) implies

$$(17) \qquad\qquad \| u(t) - \tilde{u}(t) \|^2 \leq \lambda \, \phi(t) \,.$$

On the other hand, we have

$$\phi(t) \leq \int_0^t \| u(s) - \tilde{u}(s) \| \, ds \leq \sqrt{t} \, [\int_0^t \| u(s) - \tilde{u}(s) \|^2 \, ds \,]^{\frac{1}{2}} \,,$$

by Cauchy-Schwarz inequality. Therefore, using (17):

$$(18) \qquad\qquad \phi(t)^2 \leq \lambda \, t \int_0^t \phi(s) \, ds \,.$$

The corresponding equation has a nontrivial positive solution, namely $\phi(t) = \lambda\, t^2/3$. Therefore we cannot expect (18) to imply $\phi(t) \equiv 0$ by some argument of Gronwall's Lemma type. This difficulty is due to the nature of inequality 2.4(31) which itself originates in the inequality concerning the projections onto two convex sets (Proposition 0.4.6 or [Mor 1], (2.17)): in both cases, a quadratic expression is bounded above by a linear one.

Finally, let us point out that a more direct approach to existence for the induced collisions problem could possibly have been adopted. A **discretization procedure** similar to those already presented would certainly require in this case the definition a *net* of approximate solutions, from which a converging subnet would be extracted. The practical value of such a method to the numerical analyst is doubtful.

Chapter 5

Further Applications and Related Topics

5.1. A class of second-order differential inclusions

Following Castaing [Cas 7], in this section we consider some abstract second-order differential inclusions to which some of the preceding techniques can be successfully applied in order to prove existence results. An useful concept is the following one:

Definition 1.1. A multifunction F from an open subset Ω to subsets of the Hilbert space H is said to be **anti-monotone** if, for all x and y in Ω and all $u\varepsilon F(x)$ and $v\varepsilon F(y)$:

$$(1) \qquad \lim_{h\to 0^-} \frac{\| x-y+h(u-v)\| - \| x-y\|}{h} \leq 0$$

or equivalently:

$$(2) \qquad (u-v).(x-y) \leq 0 \qquad (u\varepsilon F(x),\ v\varepsilon F(y)).$$

To see that (1) implies (2), write (limits are taken as $h\to 0^-$):

$$2(x-y).(u-v) = \lim\,[\,h\,\| u-v\|^2 + 2(x-y).(u-v)]$$

$$= \lim\, h^{-1}[\,\|(x-y)+h(u-v)\|^2 - \| x-y\|^2\,]$$

$$= \lim\,[\,\frac{\|(x-y)+h(u-v)\| - \| x-y\|}{h}\,(\,\|(x-y)+h(u-v)\| + \| x-y\|\,)\,]$$

$$\leq 0.$$

To see that (2) implies (1) — which is obvious if $x=y$ — assume that $x\neq y$ and write as above

$$\lim \frac{\|(x-y)+h(u-v)\| - \| x-y\|}{h} = \lim \frac{h\,\| u-v\|^2 + 2(x-y).(u-v)}{\|(x-y)+h(u-v)\| + \| x-y\|}$$

$$= \frac{x-y}{\| x-y\|}\cdot(u-v) \leq 0\ .$$

The main existence theorem is the following:

Theorem 1.2. [Cas 7] *Let H be a Hilbert space and $F\colon H\to 2^H$ be a multifunction with nonempty closed convex values in H. We assume that F is* *bounded, i.e., $r := \sup_{q \varepsilon H} | F(q) | = \sup \{ \, \| x \| \; : \; x \varepsilon F(q), \; q \varepsilon H \} < +\infty$ and that F is Lipschitz-continuous, with ratio λ (cf. (9)). We assume that either F is anti-monotone or H is finite-dimensional. Let $q_0 \varepsilon H$, $u_0 \varepsilon F(q_0)$ and $T > 0$. Then, there exists a Lipschitz-continuous function $u : [0, T[\to H$ such that*

(3) $\qquad u(0) = u_0 \; ;$

(4) $\qquad u(t) \varepsilon F(q(t)) \; , \; \forall t \varepsilon [0, T[\; ,$

where

$$(5) \qquad q(t) = q_0 + \int_0^t u(s) \, ds \; ,$$

and for almost every t:

(6) $\qquad - \dot{u}(t) \varepsilon N_{F(q(t))}(u(t)) \; ;$

the right-hand side of (6) *is the outward normal cone to $F(q(t))$ at $u(t)$.*

In other words, there is a solution to the Cauchy problem for the second-order differential inclusion:

$$(7) \qquad -\frac{d^2 q}{dt^2}(t) \; \varepsilon \; N_{F(q(t))}\left(\frac{dq}{dt}(t) \right) \, , \quad a.e. \; ;$$

$$(8) \qquad q(0) = q_0 \; , \; \frac{dq}{dt}(0) = u_0 \; .$$

In [Cas 7], the multifunction F is defined in an open subset of H and not in the whole Hilbert space. In order to keep the presentation as self-contained as possible, the proofs given here require less sophisticated Functional Analysis than those of Castaing. Notice that by assumption:

(9) $\qquad h(\, F(q) , F(q')) \leq \lambda \, \| q - q' \| \qquad (q, q' \varepsilon H) \, ,$

(10) $\qquad \forall q \varepsilon H \, , \; \forall u \varepsilon F(q) : \quad \| u \| \leq r \, .$

A **discretization method** is used. The time interval is partitioned by means of the points

$$t_{n,i} = i \, T 2^{-n} \quad (i = 0, \, \dots \, , \, 2^n)$$

and the approximating functions are defined by

(11) $\qquad u_n(t) = u_{n,i} \, , \; \text{if } t \varepsilon I_{n,i} := [t_{n,i} \, , \, t_{n,i+1}[\; ,$

$$(12) \qquad q_n(t) = q_0 + \int_0^t u_n(s) \, ds \, ,$$

where

$$(13) \qquad u_{n,0} = u_0 \; ; \; u_{n,i+1} = \text{proj}\left(u_{n,i} , F(q_n(t_{n,i+1})) \right).$$

Let us write $\theta_n(t) = t_{n,i}$, if t belongs to $[t_{n,i}, t_{n,i+1}[$. Then, by definition and from (10)

$$(14) \qquad u_n(t) \, \varepsilon \, F(q_n(\theta_n(t))) \, , \; \| u_n(t) \| \leq r.$$

So, all the functions q_n are Lipschitz-continuous with ratio r and they are also equibounded: $\| q_n \|_\infty \leq Q := \| q_0 \| + r\,T$.

The derivatives u_n are not Lipschitz-continuous, but the piecewise affine approximants

$$(15) \qquad v_n(t) := u_{n,i} + \frac{2^n}{T}\,(t - t_{n,i})\,(u_{n,i+1} - u_{n,i}) \, , \; \text{if } t\varepsilon \, I_{n,i} \, ,$$

are Lipschitz-continuous, with respective ratios $2^n T^{-1} \max \| u_{n,i+1} - u_{n,i} \|$. Since, by construction, $u_{n,i}$ belongs to $F(q_n(t_{n,i}))$, we have

$$\| u_{n,i+1} - u_{n,i} \| = \text{dist}\,(u_{n,i}, F(q_n(t_{n,i+1}))) \leq h(F(q_n(t_{n,i})), F(q_n(t_{n,i+1})))$$

so that, by (9) and the above remarks about q_n :

$$(16) \qquad \| u_{n,i+1} - u_{n,i} \| \leq \lambda \, \| q_n(t_{n,i}) - q_n(t_{n,i+1}) \|$$

$$\leq \lambda\,r\,| t_{n,i} - t_{n,i+1} | = \lambda\,r\,T 2^{-n} \, .$$

It follows that the v_n are equi-Lipschitz-continuous with ratio $\lambda\,r$:

$$(17) \qquad \| v_n(t) - v_n(s) \| \leq \lambda\,r\,| t - s | \, .$$

From (16) and the definitions it is also clear that $\| v_n(t) - u_n(t) \| \leq \lambda\,r\,T 2^{-n}$, hence

$$(18) \qquad \| v_n - u_n \|_\infty \to 0 \, .$$

The anti-monotonicity of the multifunction F is used to pass to the limit, by means of:

Lemma 1.3. *If F is anti-monotone, then (q_n) is a Cauchy sequence for the metric of uniform convergence.*

Proof. Denote by $w_{m,n}(t)$ or simply by $w(t)$ the quantity $\frac{1}{2} \| q_m(t) - q_n(t) \|^2$, with $m > n$. Then:

$$\frac{d^+ w}{dt}(t) = (q_m(t) - q_n(t)) \cdot (u_m(t) - u_n(t)) \qquad (t\varepsilon[0, T[) \, ,$$

since u_n is the right-derivative of q_n. If $t \varepsilon I_{n,i} \cap I_{m,j}$, then $u_m(t) = u_{m,j}$ belongs to $F(q_m(t_{m,j}))$ and $u_n(t) = u_{n,i}$ belongs to $F(q_n(t_{n,i}))$. Because F is anti-monotone (2), it follows that:

$$(q_m(t_{m,j}) - q_n(t_{n,i})) \cdot (u_m(t) - u_n(t)) \leq 0$$

and thus

$$\frac{d^+ w}{dt}(t) \leq (q_m(t) - q_m(t_{m,j})) \cdot (u_m(t) - u_n(t)) + (q_n(t_{n,i}) - q_n(t)) \cdot (u_m(t) - u_n(t)).$$

Since (14) holds and since all the q_n have Lipschitz ratio r, we have

$$\frac{d^+ w}{dt}(t) \leq r \, | t - t_{m,j} | \, 2r + r \, | t_{n,i} - t | \, 2r \leq 2 \, r^2 \, T(2^{-m} + 2^{-n}) \; .$$

Moreover, $w(0) = \frac{1}{2} \| q_m(0) - q_n(0) \|^2 = 0$. Hence $w(t) \leq 2 \, r^2 \, T(2^{-m} + 2^{-n}) \, t$ and so

$$\| q_m - q_n \|_\infty \leq 2 \, r \, T(2^{-m} + 2^{-n})^{\frac{1}{2}} \to 0 \; ,$$

when $m, \; n \to +\infty$, showing that (q_n) is a Cauchy sequence. $\square$

Before proving Theorem 1.2, let us find an equivalent formulation of the differential inclusion. In order to do so, we introduce the following definition:

Definition 1.4. A function $\phi \colon [0, T] \times \dot H \to H$ is called an *F-selection* if it is measurable, bounded, continuous with respect to the H variable and such that

$$(19) \qquad\qquad \forall t \varepsilon [0, T[\, , \; \forall q \varepsilon H : \; \phi(t, q) \, \varepsilon \, F(q) \; .$$

Proposition 1.5. *A Lipschitz-continuous function $u \colon [0, T[\to H$ which satisfies conditions (3), (4) and (5) is a solution to the differential inclusion (6) if and only if*

$$(20) \qquad\qquad \int_0^T [\phi(t, q(t)) - u(t)] \cdot \dot u(t) \, dt \geq 0 \; ,$$

for every F-selection ϕ.

Proof. 1) Let u be a solution and ϕ be an F-selection. Since a.e. $-\dot u(t)$ belongs to the outward normal cone to $F(q(t))$ at $u(t)$ and since $\phi(t, q(t)) \varepsilon F(q(t))$, it turns out that

$$(\phi(t, q(t)) - u(t)) \cdot \dot u(t) \geq 0 \; , \; \text{a.e.}$$

whence (20).

2) Assume that (3)–(5) and (20) hold. We prove (6).

Fix t in $[0, T[$. For every $z\varepsilon\, F(q(t))$ and every small positive ϵ, we define the F-selection

$$\phi(s, q) = \begin{cases} \text{proj}(z, F(q)) & \text{if } s\varepsilon[t, t+\epsilon] \\ \text{proj}(u(s), F(q)) & \text{otherwise .} \end{cases}$$

Then $\phi(s, q(s)) = u(s)$ if $s \notin [t, t+\epsilon]$ and (20) implies

$$\int_t^{t+\epsilon} [\text{proj}(z, F(q(s))) - u(s)] \cdot \dot{u}(s)\, ds \geq 0\ .$$

Dividing by ϵ and letting ϵ go to zero, we obtain, for almost every t,

$$[\text{proj}(z, F(q(t))) - u(t)] \cdot \dot{u}(t) \geq 0\,,$$

i.e. :
$$(z - u(t)) \cdot \dot{u}(t) \geq 0\quad,\ \text{a. e. .}$$

Since $u(t)\varepsilon\, F(q(t))$ and z is arbitrary in $F(q(t))$, we may conclude that (6) holds:

$$-\dot{u}(t)\,\varepsilon\, N_{F(q(t))}(u(t))\ .\qquad\qquad\square$$

The first existence result (Theorem 1.2) is proved next.

Proof of Theorem 1.2. Consider the approximating sequences (q_n), $(\dot{q}_n) = (u_n)$ and (v_n) defined above.

If F is anti-monotone, we have seen (Lemma 1.3) that (q_n) is a Cauchy sequence for the norm of uniform convergence, hence it converges uniformly to a Lipschitz-continuous function q.

If H is finite-dimensional, we use instead Ascoli-Arzelà's theorem. Since (q_n) is a sequence of equi-Lipschitz-continuous functions with the same initial value, then we may extract from (q_n) a sequence that converges uniformly to a Lipschitz-continuous function q.

Since (v_n) and (u_n) are uniformly bounded in norm and in variation and since (18) holds, we may suppose (if necessary by taking a subsequence) that both converge pointwisely weakly to some function u. As every v_n has Lipschitz ratio $\lambda\, r$ ((17)), we get, using the weak lower semicontinuity of the norm:

$$\| u(t) - u(s) \| \leq \liminf \| v_n(t) - v_n(s) \| \leq \lambda r\, | t - s |\,;$$

that is, u is Lipschitz-continuous with ratio $\lambda\, r$.

1. Obviously, $u(0) = \text{w-lim } u_n(0) = u_0$.

2. For every $h \varepsilon H$, $q_n \cdot h \to q \cdot h$ uniformly, so $u_n \cdot h = \dot{q}_n \cdot h \to \dot{q} \cdot h$ at least in the sense of distributions; on the other hand, $u_n \cdot h \to u \cdot h$ pointwisely, so $u = \dot{q}$ almost everywhere and (5) holds.

3. By (9) and (14),

$$\text{dist}\left(u_n(t), F(q(t))\right) \leq h\left(F(q_n(\theta_n(t))), F(q(t))\right) \leq \lambda \parallel q_n(\theta_n(t)) - q(t) \parallel ,$$

which converges to zero, because $\theta_n(t) \to t$ and $q_n \to q$ uniformly. Recalling that $u_n(t)$ converges weakly to $u(t)$ and that $F(q(t))$ is a closed convex set, we see that (4) holds:

$$u(t) \; \varepsilon \; F(q(t)) \; .$$

4. In order to prove (6), we use Proposition 1.5. Let us consider an F-selection ϕ and show that (20) holds. Define $\sigma_n(t) := t_{n,i+1} \quad$ if $t \varepsilon [t_{n,i}, t_{n,i+1}[$. As

$$\dot{v}_n(t) = 2^n \, T^{-1} \left(u_{n,i+1} - u_{n,i}\right) = 2^n \, T^{-1} \left[\,\text{proj}\left(u_{n,i}, F(q_n(t_{n,i+1}))\right) - u_{n,i}\right],$$

it is clear that $-\dot{v}_n(t)$ belongs a.e. to the outward normal cone:

$$N_{F(q_n(t_{n,i+1}))}(u_{n,i}) = N_{F(q_n(\sigma_n(t)))}(u_n(t)) \; .$$

By definition of this cone and since

$$\phi_n(t) := \phi(t, q_n(\sigma_n(t))) \; \varepsilon \; F(q_n(\sigma_n(t))) \; ,$$

we have $\dot{v}_n(t) \cdot [\phi_n(t) - u_n(t)] \geq 0$ a.e. and so:

$$(21) \qquad \int_0^T \dot{v}_n(t) \cdot \phi_n(t) \, dt \; \geq \; \int_0^T \dot{v}_n(t) \cdot u_n(t) \, dt \, .$$

Since $(\dot{v}_n)$ is bounded in $L^\infty(I, H)$, we may assume that it converges weakly-$*$ to a function which actually is $\dot{u}$. Moreover, the assumptions on ϕ and the convergence $q_n(\sigma_n(t)) \to q(t)$ imply that $\phi_n(t) \to \phi(t, q(t)) = \hat{\phi}(t)$ in L^1 norm. It follows that

$$\left| \int_0^T \dot{v}_n \cdot (\phi_n - \hat{\phi}) \right| \; \leq \; \lambda \, r \int_0^T \parallel \phi_n - \hat{\phi} \parallel \to 0 \, , \qquad \int_0^T (\dot{v}_n - \dot{u}) \cdot \hat{\phi} \to 0$$

and so:

$$\int_0^T \dot{v}_n(t) \cdot \phi_n(t) \, dt \; \to \; \int_0^T \dot{u}(t) \cdot \phi(t, q(t)) \, dt \, .$$

Concerning the right-hand side of (21), we note that $u_n - v_n \to 0$ uniformly, $v_n(t) \to u(t)$ weakly and $v_n(0) = u_0 = u(0)$ imply that:

$$\liminf \int_0^T \dot{v}_n \cdot u_n \, dt = \liminf \int_0^T \dot{v}_n \cdot v_n \, dt = \liminf \tfrac{1}{2} \left(\| v_n(T) \|^2 - \| v_n(0) \|^2 \right)$$

$$\geq \tfrac{1}{2} \| u(T) \|^2 - \tfrac{1}{2} \| u(0) \|^2 = \int_0^T u(t) \cdot \dot{u}(t) \, dt .$$

Thus, (21) implies

$$\int_0^T \dot{u}(t) \cdot \phi(t, q(t)) \, dt \geq \int_0^T u(t) \cdot \dot{u}(t) \, dt ,$$

that is to say (20) . The proof is finished. ◻

With standard changes in the reasoning, a local existence result is obtained for multifunctions defined not in the whole space H, but only in some open subset of it (see [Cas 7]).

We give another existence theorem, where the assumptions (about the interior) and the techniques remind us of Chapter 2:

Theorem 1.6. [Cas 7] *Let F be a Hausdorff-continuous multifunction from the Hilbert space H to closed convex subsets with nonempty interior in H. Assume that $\overline{B}(a, \rho) \subset F(q) \subset \overline{B}(0, r)$, $\forall q \varepsilon H$, and that $q_0 \varepsilon H$, $u_0 \varepsilon F(q_0)$ and $T > 0$. Suppose that one of the following conditions is satisfied:*

(a) H is finite-dimensional; or

(b) F is anti-monotone.

Then, there exists a continuous function of bounded variation $u: [0, T[\to H$ such that (3) $u(0) = u_0$ and (4) $u(t) \varepsilon F(q(t))$,$\forall t \varepsilon [0, T[$ hold, where (5) $q(t) = q_0 + \int_0^t u(s) \, ds$, and for $|\, du \,|$ -almost every t:

$$(22) \qquad\qquad -\frac{du}{|\, du \,|}(t) \, \varepsilon \, N_{F(q(t))}(u(t)) .$$

Proof. We define the approximants $(q_n), (u_n)$ and (v_n) as before. Then $\| u_n(t) \| \leq r$ and the q_n are Lipschitz-continuous with constant r and uniformly bounded by $\| q_0 \| + r\,T$. The total variation of u_n (or v_n) is estimated by means of Lemma 0.4.4: since all the sets $F(q)$ contain $\overline{B}(a, \rho)$,

$$(23) \qquad \operatorname{var} u_n = \sum_{0 \leq i < 2^n} \| u_{n,i+1} - u_{n,i} \|$$

$$= \sum_{0 \leq i < 2^n} \| \operatorname{proj}(u_{n,i}, F(q_n(t_{n,i+1}))) - u_{n,i} \|$$

$$\leq C := \tfrac{1}{2\rho} \| u_0 - a \|^2 .$$

If H is finite-dimensional, then by Theorem 0.2.1, Ascoli-Arzelà's theorem and Lebesgue's dominated convergence theorem, we may extract a subsequence, still denoted by (u_n), which converges pointwisely to a bv function u (whence $u(0) = u_0$) and such that $q_n(t) = q_0 + \int_0^t u_n(s)\, ds$ converges uniformly to a Lipschitz-continuous function q, given by (5). Thus, we have $u_n(t) \in F(q_n(\theta_n(t)))$, $u_n(t) \to u(t)$ and also $q_n(\theta_n(t)) \to q(t)$, because

$$\| q_n(\theta_n(t)) - q(t) \| \leq \| q_n(\theta_n(t)) - q_n(t) \| + \| q_n(t) - q(t) \|$$

$$\leq r \, | \theta_n(t) - t | + \| q_n(t) - q(t) \| \to 0 \, .$$

These imply (4), because F is continuous.

If H is infinite-dimensional and F is anti-monotone, then Lemma 1.3 yields that (q_n) converges uniformly to a Lipschitz-continuous function $q : [0, T] \to H$ with ratio r. Extracting a subsequence as in Theorem 0.2.1, we may assume that $u_n = \dot q_n$ converges pointwisely weakly to a bv function u, which again satisfies (3) and (5). Since $u_n(t)$ converges weakly to $u(t)$, $q_n(\theta_n(t)) \to q(t)$ and F is continuous with weakly closed values, then $u(t) \in F(q(t))$ still holds.

We prove below that in fact (u_n) converges uniformly to u and that u is continuous. Thus, as the multifunction $t \to F(q(t))$ is continuous with closed convex values having nonempty interior, Lemma 2.2.4 applies: the differential inclusion (22) is equivalent to

$$(24) \qquad z . (u(t) - u(s)) \geq \tfrac{1}{2} \| u(t) \|^2 - \tfrac{1}{2} \| u(s) \|^2 \, ,$$

for every $s < t$ and $z \in F(q(\tau))$, $\forall \tau \in [s, t]$.

By continuity of u, we may assume that s and t are multiples of $2^{-n} T$ for n large enough, so that $u_n(\theta_n(s)) = u_n(s) \to u(s)$ and similarly $u_n(\theta_n(t)) \to u(t)$. Notice also that $\operatorname{dist}(z, F(q_n(\tau))) \leq h(F(q(\tau)), F(q_n(\tau))) \to 0$, uniformly; hence the sequence of functions

$$\phi_n(\tau) := \operatorname{proj}(z, F(q_n(\tau))) \qquad (\tau \in [s, t])$$

converges uniformly to z. For the discretization nodes contained in $[s, t]$, we have

$$(u_{n,i+1} - \phi_n(t_{n,i+1})) . (u_{n,i+1} - u_{n,i}) \leq 0 \, ,$$

since $\phi_n(t_{n,i+1})$ belongs to $F(q_n(t_{n,i+1}))$ and $u_{n,i+1}$ is the projection of $u_{n,i}$ in this set. Hence:

$$z \cdot (u_n(t) - u_n(s)) = z \cdot \sum (u_{n,i+1} - u_{n,i})$$

$$\geq \sum (z - \phi_n(t_{n,i+1})) \cdot (u_{n,i+1} - u_{n,i}) + \sum u_{n,i+1} \cdot (u_{n,i+1} - u_{n,i}),$$

so that, by (23) and §2.3(20):

$$z \cdot (u_n(t) - u_n(s)) \geq - \sup_\tau \| z - \phi_n(\tau) \| \sum \| u_{n,i+1} - u_{n,i} \|$$

$$+ \tfrac{1}{2} \sum (\| u_{n,i+1} \|^2 - \| u_{n,i} \|^2)$$

$$\geq - C \| z - \phi_n \|_\infty + \tfrac{1}{2} \| u_n(t) \|^2 - \tfrac{1}{2} \| u_n(s) \|^2 .$$

Passing to the limit as $n \to \infty$, we obtain (24) as needed.

In order to complete the proof, let us first note that, if (u_n) converges uniformly to u, then so does the sequence of continuous functions (v_n), hence u is continuous. In fact, we have:

$$(25) \qquad \| v_n - u_n \|_\infty = \max_i \| u_{n,i+1} - u_{n,i} \|$$

$$\leq \max_i h(F(q_n(t_{n,i})) , F(q_n(t_{n,i+1}))) \to 0 .$$

Given $\epsilon > 0$, for every t we take $\delta(t) > 0$ such that $\| x - q(t) \| < \delta(t)$ implies $h(F(x), F(q(t))) < \epsilon/2$. The set $q([0,T])$ is compact, so there is a finite open covering, denoted by $\Omega = B(q(t_1), \delta(t_1)/2) \cup ... \cup B(q(t_p), \delta(t_p)/2)$. Since (q_n) converges uniformly to q, then $2^{-n} T < \min \delta(t_k)/2r$ and $q_n(t) \in \Omega$ for all t and for large enough n. Thus, for every i, there is t_k such that $\| q_n(t_{n,i}) - q(t_k) \| < \delta(t_k)/2$, while $\| q_n(t_{n,i+1}) - q_n(t_{n,i}) \| \leq r 2^{-n} T < \delta(t_k)/2$. So, $\| q_n(t_{n,i+1}) - q(t_k) \| < \delta(t_k)$. Hence, for large enough n,

$$h(F(q_n(t_{n,i+1})), F(q_n(t_{n,i}))) \leq h(F(q_n(t_{n,i+1})), F(q(t_k))) + h(F(q(t_k)), F(q_n(t_{n,i})))$$

$$< \epsilon .$$

This proves (25). The same argument ensures that, if m and n $(m > n)$ are large enough, then

$$(26) \qquad\qquad h(F(q_n(t_{n,j})) , F(q_m(t_{m,i}))) < \epsilon ,$$

for every i and j with $I_{n,j} \cap I_{m,i} \neq \emptyset$.

Let $t \in [t_{m,i}, t_{m,i+1}[\cap [t_{n,j}, t_{n,j+1}[$. We have

$$u_m(t) = u_{m,i} = \mathrm{proj}\, (u_m(t_{m,i-1}), F(q_m(t_{m,i}))) \quad \text{and}$$

$$u_n(t) = u_{n,j} = \mathrm{proj}\, (u_{n,j-1}, F(q_n(t_{n,j}))) = \mathrm{proj}\, (u_n(t_{m,i-1}), F(q_n(t_{n,j}))) ,$$

since either $t_{m,i} = t_{n,j}$ and $u_n(t_{m,i-1}) = u_{n,j-1}$ or $t_{m,i} > t_{n,j}$ and $u_n(t_{m,i-1}) = u_{n,j}$. Taking (26) into account, the inequality of Moreau about projections into convex sets (Proposition 0.4.7) yields :

$$\| u_m(t) - u_n(t) \|^2 \leq \| u_m(t_{m,i-1}) - u_n(t_{m,i-1}) \|^2$$
$$+ 2\epsilon [\| u_m(t) - u_m(t_{m,i-1}) \| + \| u_n(t) - u_n(t_{m,i-1}) \|].$$

By induction, we easily see that:

$$\| u_m(t) - u_n(t) \|^2 \leq 2\epsilon [\text{var}(u_m; 0, t) + \text{var}(u_n; 0, t)] \leq 4\,C\epsilon .$$

Hence $\| u_m - u_n \|_\infty \to 0$, as $m, n \to \infty$ and (u_n) converges uniformly. $\square$

5.2. Lipschitz approximations of sweeping processes

Our purpose here is to give an overview of works by Valadier [Val 1-3] and by Gavioli [Gav] on approximations of multifunctions from the interior and from the exterior, respectively, by more regular ones (e.g. Lipschitz). These approximations are used to give new proofs of the existence of a solution to a sweeping process. *En passant*, we shall need some mathematical tools, unused in the foregoing; in this context, a reference to [Cas 8] should be made.

For completeness, let us prove existence for the Lipschitz case. The result is due to Moreau [Mor 2, 9] and the proof is that of [Cas-Val] Theorem VII-19 or [Val 2] Théorème 2.

Theorem 2.1. *Let H be a (separable) Hilbert space, let $C: [0, T] \to 2^H$ be a multifunction with nonempty closed convex values and $a \varepsilon\, C(0)$. Assume that C is Lipschitz-continuous for the Hausdorff distance:*

$$h(C(t), C(s)) \leq k \,|\, t - s \,|, \quad \forall\, t, s \,\varepsilon\, [0, T] .$$

Then there is a unique Lipschitz-continuous function $u: [0, T] \to H$ such that:

(1) $\qquad u(0) = a$;

(2) $\qquad \forall t, \ u(t) \,\varepsilon\, C(t)$;

(3) $\qquad -\dfrac{du}{|\,du\,|}(t) \,\varepsilon\, N_{C(t)}(u(t))$, *a. e.* t .

Moreover, we have:

(4) $\qquad\qquad\qquad \| \dfrac{du}{dt} \| \leq k$, *a.e.*

and if, for some $z \varepsilon\, H$ and $r > 0$, $C(t) \supset \overline{B}(z, r) \,\forall t$, then

$$(5) \qquad \mathrm{var}\,(u; 0, T) = \int \left| \frac{du}{dt} \right|\, dt \ \leq\ k(r, \| a - z \|) \ \leq\ \frac{1}{2r}\, \| a - z \|^2\,.$$

Proof. Uniqueness is a consequence of the monotonicity of $x \to N_{C(t)}(x)$.

For existence, assume with no loss of generality that $T = 1$ (to simplify some expressions). For $n \geq 1$, define the polygonal function x_n first on the set D_n of points $i 2^{-n}$ by

$$x_n(0) = a, \quad x_n(i 2^{-n}) = \mathrm{proj}\,(x_n((i-1)\, 2^{-n}),\, C(i 2^{-n}))\,,$$

and then on the intervals $](i-1)\, 2^{-n}, i 2^{-n}[$ by affine interpolation. Denote $\sigma_n(0) = 0$, $\sigma_n(t) = i 2^{-n}$ if $t \, \varepsilon \,](i-1)\, 2^{-n}, i 2^{-n}]$. In each one of the above open intervals, the derivative x_n' is constant:

$$x_n'(t) = 2^n [\, x_n(i 2^{-n}) - x_n((i-1)\, 2^{-n})]$$

and we have

$$-x_n'(t) \ \varepsilon \ N_{C(i 2^{-n})}(x_n(i 2^{-n})) = N_{C(\sigma_n(t))}(x_n(\sigma_n(t)))\,,$$

$$\| x_n'(t) \| = 2^n \, \mathrm{dist}\,(x_n((i-1)\, 2^{-n}),\, C(i 2^{-n})) \ \leq\ 2^n \, h(\, C((i-1)\, 2^{-n}),\, C(i 2^{-n}))$$

$$\leq\ 2^n k 2^{-n} = k\,.$$

This allows the extraction of a subsequence (u_p') from (x_n') such that the sequence of derivatives (u_p') converges $\sigma(L_H^\infty, L_H^1)$ to $u' \varepsilon L^\infty(0,1; H)$ with $\| u'(t) \| \leq k$ a.e.. The correspondent subsequence of (x_n) is denoted by (u_p). Let

$$u(t) = a + \int_0^t u'(s)\, ds.$$

If t belongs to $D = \bigcup_n D_n$, then $u_p(t) \, \varepsilon \, C(t)$ for every p large enough and, since $u_p(t) \to u(t)$ weakly, we have $u(t) \varepsilon C(t)$. Since u and C are continuous and since D is dense in $[0, 1]$, we readily obtain property (2).

Denote by $\delta(x \,|\, C)$ the *indicator function* of a convex set C (which is zero on C and $+\infty$ outside) and by $\delta^*(x' \,|\, C)$ its support function $(= \sup \{x' . x \,|\, x \varepsilon \, C\})$. Consider the functionals

$$I(v) = \int_0^1 \delta^*(v(t) \,|\, C(t))\, dt \quad (v \varepsilon \, L^\infty(0, 1; H)),$$

$$I^*(w) = \int_0^1 \delta(w(t) \,|\, C(t))\, dt \quad (w \varepsilon \, L^1(0, 1; H)).$$

According to [Roc 1] (for separable H) or [Cas-Val] (Theorem VII-14) these are mutually polar (conjugate) functionals, hence they are both weakly lower semicontinuous.

If $z \in L^\infty$, then $\int_0^1 z(t) \cdot (\int_0^t z(s)\, ds)\, dt = \frac{1}{2} \| \int_0^1 z(t)\, dt \|^2$, by Fubini's

theorem. Thus, the functional

$$z \to q(z) := \int_0^1 z(t) \cdot a\, dt + \frac{1}{2} \| \int_0^1 z(t)\, dt \|^2$$

is weakly lower semicontinuous in L^∞ and satisfies $q(u_p') = \int_0^1 u_p'(t) \cdot u_p(t)\, dt$
$= \langle u_p' , u_p \rangle$.

Let $n = n_p$ be such that $u_p' = x_{n_p}'$. We recall that (see e. g. [Cas-Val]
Theorem II-18):

$$| \delta^*(x' \mid C) - \delta^*(x' \mid \tilde{C}) | \leq \| x' \| \, h(C, \tilde{C}) \, .$$

By the assumptions on C, we have:

$$| \int_0^1 \delta^*(-u_p'(t) \mid C(t))\, dt - \int_0^1 \delta^*(-u_p'(t) \mid C(\sigma_n(t)))\, dt | \leq$$

$$\leq k^2 \int_0^1 | \sigma_n(t) - t |\, dt \, ;$$

again using $\| u_p'(t) \| \leq k$ a.e. , we also have:

$$| \int_0^1 u_p'(t) \cdot u_p(t)\, dt - \int_0^1 u_p'(t) \cdot u_p(\sigma_n(t))\, dt | \leq k^2 \int_0^1 | \sigma_n(t) - t |\, dt \, .$$

Moreover, in the open subintervals, $-u_p'(t)$ belongs to the outward normal
cone to $C(\sigma_n(t))$ at $u_p(\sigma_n(t)) = x_n(\sigma_n(t))$ (where $n = n_p$) and this can be
expressed also by the geometrically clear condition

$$\delta^*(-u_p'(t) \mid C(\sigma_n(t))) = -u_p'(t) \cdot u_p(\sigma_n(t)) \, .$$

From the above considerations, it follows that:

$$I(-u_p') + \langle u_p' , u_p \rangle \leq 2 k^2 \int_0^1 | \sigma_n(t) - t |\, dt = 2^{-n} k^2 \, ,$$

whence, by the lower semicontinuity of I and q:

$$I(-u') + \langle u' , u \rangle = \int_0^1 [\delta^*(-u'(t) \mid C(t)) + u'(t) \cdot u(t)]\, dt \leq 0 \, .$$

But (2) and the definition of the support function imply that
$\delta^*(-u'(t) \mid C(t)) \geq -u'(t) \cdot u(t)$. Hence, we must have:

$$\delta^*(-u'(t) \mid C(t)) = -u'(t) \cdot u(t) \quad \text{a.e.} \, ,$$

which, taken together with (2), means that (3) holds (by a well-known result of Convex Analysis).

Finally, assume that $C(t) \supset \overline{B}(z,r)$, $\forall t$. Then, by Lemma 0.4.4:

$$\text{var } u_p = \text{var } x_{n_p} = \sum \| x_{n_p}(i 2^{-n}) - x_{n_p}((i-1) 2^{-n}) \|$$

$$\leq l(r, \| a-z \|) \leq \frac{1}{2r} \| a-z \|^2 ,$$

so that, in the limit, we get estimate (5). $\square$

The next result is a slightly improved version of the incomplete characterization of the solution to lsc sweeping processes given in Lemma 2.2.4. Here the assumption on the interior is dropped.

Proposition 2.2. [Val 1](Prop. 6) *Let C be a lower semicontinuous multifunction from $[0, T]$ to nonempty closed convex subsets of a Hilbert space H and let $u:[0, T] \to H$ be an rcbv function.*

If u is a solution to the sweeping process by C, then

$$(6) \qquad \forall s < t, \ \int_{]s, \, t]} \phi \cdot du \geq \tfrac{1}{2} (\| u(t) \|^2 - \| u(s) \|^2) ,$$

$\forall \phi$ continuous selection of C.

Conversely, if u is a selection of C and (6) is satisfied, then

$$- \frac{du}{|\,du\,|}(t) \ \varepsilon \ N_{C(t)}(u(t))$$

holds at $|\,du\,|$ -almost every continuity point t of u.

Proof. If u is a solution, then $\phi(\tau) \, \varepsilon \, C(\tau)$ and so the condition on the density implies

$$[\phi(\tau) - u(\tau)] \cdot \frac{du}{|\,du\,|}(\tau) \geq 0 ,$$

for $|\,du\,|$ -almost every τ. Integrating over $]s, t]$ with respect to $|\,du\,|$ and taking into account the right-continuity of u, we obtain (6).

Suppose that u is an rcbv selection selection of C satisfying (6). Let N be a $|\,du\,|$ -null set such that Jeffery's formulas for the densities $du/|\,du\,|$ and $d(\| u \|^2)/|\,du\,|$ hold for $t \notin N$. If T is a continuity point of u, then we add it to N. Let $t\varepsilon [0, T[\setminus N$ be a continuity point of u not belonging to N. Let $x \varepsilon \, C(t)$. Since C is lower semicontinuous, by Michael's selection theorem [Mic] there exists ϕ, a continuous selection of C, such that $\phi(t) = x$. If

$0 < \epsilon \leq T - t$, then, since $\{t\}$ has zero measure and by (6), we have:

$$(7) \qquad \int_{[t,t+\epsilon]} \phi \cdot du \;\geq\; v(t+\epsilon) - v(t) = dv([t, t+\epsilon]) ,$$

where $v(t) := \frac{1}{2} \| u(t) \|^2$. But

$$\int_{[t,t+\epsilon]} \phi \cdot du = x \cdot du([t, t+\epsilon]) + \int_{[t,t+\epsilon]} (\phi - x) \cdot du$$

and

$$\frac{| \int_{[t,t+\epsilon]} (\phi - x) \cdot du |}{| du | ([t, t+\epsilon])} \;\leq\; \sup \{ \| \phi(\tau) - x \| \; : \; \tau \in [t, t+\epsilon] \} \;\to\; 0 ,$$

as $\epsilon \to 0$. Hence, dividing (7) by $| du | ([t, t+\epsilon])$ and letting ϵ go to zero, we get

$$x \cdot \frac{du}{| du |}(t) \;\geq\; \frac{dv}{| du |}(t) = \frac{u^+(t) + u^-(t)}{2} \cdot \frac{du}{| du |}(t) = u(t) \cdot \frac{du}{| du |}(t)$$

and, as $x \in C(t)$ is arbitrary, the result follows. $\qquad\qquad\square$

The converse does not hold if u is discontinuous at t, as shown by the following (counter)example.

Example 2.3. [Val 1] Take $C(t) = [0, +\infty[$, if $t \in [0, 1[$ and $C(t) = [1, +\infty[$ if $t \in [1, 2]$. Then $u = \sqrt{2} \, \chi_{[1,2]}$ is not a solution to the sweeping process by C: in fact, $u^-(1) = 0$ but $u(1) = \sqrt{2} \neq \mathrm{proj}\,(0, C(1)) = 1$. Nevertheless, u satisfies (6). Indeed, if $1 \notin]s, t]$, then u is constant in that interval and (6) is obvious; while if $1 \in]s, t]$ then $du = \sqrt{2}\, \delta_1$ (Dirac measure) and $\phi(1) \geq 1$ so that

$$\frac{1}{2}(u(t)^2 - u(s)^2) = 1 \;\leq\; \sqrt{2}\, \phi(1) = \int_{]s,t]} \phi \cdot du .$$

The approximation by Lipschitz multifunctions is the object of the next theorem (for more results of this type and an interesting discussion see [Val 3] and references therein). We call *n-Lipschitz selector* of C any selection of C such that $\| \phi(s) - \phi(t) \| \leq n \, | s - t |$.

Theorem 2.4. [Val 1] *Let C be a lower semicontinuous multifunction from $[0, T]$ to closed convex subsets of $\mathbb{R}^d$ such that, for all t, $C(t) \supset B(z, r)$ ($z \in \mathbb{R}^d$, $r > 0$ fixed). For $n \in \mathbb{N}$, define:*

$$(8) \qquad C_n(t) = \{ f(t) : f \text{ is a } n\text{-Lipschitz selector of } C \} .$$

Then C_n is a n-Lipschitz multifunction ($h(C_n(t), C_n(s)) \leq n \, | t - s | $) with nonempty closed convex values and the increasing sequence (C_n) approximates

C in the following sense:

(9) $\forall t,\ \ \text{int } C(t) \subset \bigcup_{n \in \mathbf{N}} C_n(t) \subset C(t)\ ;$

in particular, $C(t)$ is the closure of the union of all the $C_n(t)$.

Proof. It is obvious that $z \varepsilon\, C_n(t) \subset C(t)$ and that $C_n(t) \subset C_{n+1}(t)$.

Since a convex combination of n-Lipschitz selectors of C is still a n-Lipschitz selector, we see that $C_n(t)$ is a convex set. If x is the limit of $(f_k(t))$, where (f_k) is a sequence of n-Lipschitz selectors, then extracting (by Ascoli-Arzelà's theorem) a subsequence converging uniformly to some function f and observing that f is a n-Lipschitz selector of C and $x = f(t)$, we prove that $C_n(t)$ is a closed set.

To show that C_n is a n-Lipschitz multifunction, take an arbitrary $x = f(t) \varepsilon\, C_n(t)$ and, since $\text{dist}\,(x, C_n(s)) \leq \| f(t) - f(s) \| \leq n\,|\,t-s\,|$, conclude that $e(C_n(t), C_n(s)) \leq n\,|\,t-s\,|$ and, by symmetry, $h(C_n(t), C_n(s)) \leq n\,|\,t-s\,|$.

It remains to prove that if x is in the interior of $C(t)$ then it belongs to some $C_n(t)$. Because C is lower semicontinuous, by Lemma 2.4.2 there is a neighbourhood V of t such that x belongs to the interior of $C(s)$, for all s in V. Let $\alpha : [0, T] \to [0,1]$ be a Lipschitz-continuous function whose support is contained in V, with $\alpha(t) = 1$. Then $f(s) = \alpha(s)\,x + (1 - \alpha(s))\,z$ is a Lipschitz-selector of C and $f(t) = x$. $\square$

We shall also need to approach uniformly a selection of C by selections of the approximate multifunctions C_n.

Proposition 2.5. [Cas 8] (Prop. 2.5) *Let (C_n) be an increasing sequence of continuous multifunctions from $[0, T]$ to closed convex subsets of $\mathbf{R}^d$, converging to a multifunction C in the following sense:*

$$C(t) = \text{cl}\,(\bigcup_n C_n(t))\ ,\ \forall t\ .$$

If ϕ is a continuous selection of C, then there exists a sequence (ϕ_n) which converges uniformly to ϕ and such that, for every n, ϕ_n is a continuous selection of C_n .

Proof. Define $r_n(t) = \text{dist}\,(\phi(t), C_n(t))$. These are continuous functions (ϕ and C_n are continuous). For every t, $r_n(t)$ is a nonincreasing sequence (because

$C_n(t) \subset C_{n+1}(t)$) and it converges to zero, since $\phi(t)$ belongs to the closure of the union $\bigcup_n C_n(t)$. By Dini's theorem, $r_n(t) \to 0$ uniformly in $[0, T]$.

Consider the multifunctions with nonempty closed convex values defined by:

$$\Gamma_n(t) = \mathrm{cl}\, [C_n(t) \cap B(\phi(t), r_n(t) + \tfrac{1}{n})] = C_n(t) \cap \overline{B}(\phi(t), r_n(t) + \tfrac{1}{n}) \ .$$

They are lower semicontinuous (this is shown first for the set inside the brackets and then for its closure). Thus, by Michael's selection theorem [Mic], there is a continuous selection $\phi_n : [0, T] \to \mathbb{R}^d$, $\phi_n(t) \varepsilon \Gamma_n(t) \subset C_n(t)$. By construction, $\| \phi_n - \phi \|_\infty \leq \| r_n \|_\infty + \tfrac{1}{n} \to 0$. $\qquad \square$

We can now prove a convergence result for solutions to sweeping processes, as in [Val 1], Theorem 11, profiting from some ideas presented in Chapter 2 and in [Cas 8], Théorème 2.3. Combining this convergence result with Theorem 2.4, we solve the sweeping process by lower semicontinuous multifunctions with left-closed graph and whose values have nonempty interior.

Theorem 2.6. [Val 1] *Let (C_n) be an increasing sequence of Lipschitz multifunctions from $[0, T]$ to closed convex subsets of $\mathbb{R}^d$, containing a fixed ball $\overline{B}(z, r)$. Suppose that the "limit" multifunction $C(t) = \mathrm{cl}\,(\bigcup_n C_n(t))$, which is lower semicontinuous, has a left-closed graph (that is, closed with respect to the left topology in $[0, T]$ and the usual one in $\mathbb{R}^d$). Let $a \varepsilon \, C(0)$ and $a_n \varepsilon \, C_n(0)$, with $a_n \to a$.*
Then, the sequence (u_n) of solutions to the sweeping processes by C_n with $u_n(0) = a_n$ converges pointwisely to the unique solution u of the sweeping process by C with initial value $u(0) = a$.

Proof. By Theorem 2.1, the Lipschitz-continuous functions (u_n) have uniformly bounded variations: $\mathrm{var}\, u_n \leq \tfrac{1}{2r} \sup_n \| a_n - z \|^2 = c_1$, and are uniformly bounded: $\| u_n \|_\infty \leq \sup_n \| a_n \| + c_1 = c_2$. Hence, (u_n) is relatively compact for the topology of pointwise convergence. Any sublimit function u is a bv function ($\mathrm{var}\, u \leq c_1$) with $u(0) = a$ and $\forall t$, $u(t) \varepsilon \, C(t)$ (since for every n, $u_n(t) \varepsilon \, C_n(t) \subset C(t)$). Moreover, the compactness theorems in §0.2 ensure that some *usual* subsequence, still denoted by (u_n), converges pointwisely to u (see also [Val 2], Lemma 8, for a refinement).

1)   We show that u is right-continuous at every $t \varepsilon [0, T[$.

Let $\epsilon > 0$. Since $u(t) \varepsilon C(t)$, by Lemma 2.4.4 there is some $x_0 \varepsilon C(t)$ and some $r_0 > 0$ such that $l(r_0, \| u(t) - x_0 \|) \leq \frac{1}{2 r_0} \| u(t) - x_0 \|^2 < \epsilon$ and $\overline{B}(x_0, r_0) \subset \text{int } C(t)$. The multifunction $n \to C_n(t)$, if extended to $n = \infty$ by $C_\infty(t) = C(t)$, is lower semicontinuous at infinity. Then, Lemma 2.4.2 implies that, for large n, $C_n(t)$ contains the closed ball $\overline{B}(x_0, r_0)$ in its interior; moreover, we may ensure that $l(r_0, \| u_n(t) - x_0 \|) < \epsilon$ holds for $n \geq N$. By the lower semicontinuity of C_N and Lemma 2.4.2, take $\delta > 0$ in such a way that $C_N(\tau) \supset \overline{B}(x_0, r_0)$, $\forall \tau \varepsilon [t, t+\delta]$. This still holds with N replaced by $n \geq N$, so that the estimate in Theorem 2.1 implies:

$$\text{var}\,(u_n; t, t+\delta) \leq l(r_0, \| u_n(t) - x_0 \|) < \epsilon.$$

It follows that

$$\forall \tau \varepsilon [t, t+\delta], \; \| u(\tau) - u(t) \| \leq \text{var}\,(u; t, t+\delta) \leq \epsilon .$$

2)   We show that u is left-continuous at every $t \varepsilon\,]0, T]$.

Taking t_n converging to t from the left, we see that $(t_n, u(t_n)) \varepsilon$ graph C converges to $(t, u^-(t))$; since the graph of C is left-closed, we then conclude that $u^-(t) \varepsilon C(t)$. Let $\epsilon > 0$. The arguments in 1), if applied to $u^-(t)$ and to an interval to the left of t, show that there exists a ball $\overline{B}(x_1, r_1)$ such that $l(r_1, \| u^-(t) - x_1 \|) < \epsilon$ and that, for large n, $\overline{B}(x_1, r_1) \subset C_n(\tau)$, $\forall \tau \varepsilon [t', t]$ where $t' < t$. Moreover, by taking larger t' and n if necessary, we have $\| u(t') - u^-(t) \| < \epsilon$, $l(r_1, \| u(t') - x_1 \|) < \epsilon$, $\| u_n(t') - u(t') \| < \epsilon$ and so (Theorem 2.1) $\| u_n(t) - u_n(t') \| \leq \text{var}\,(u_n; t, t') \leq l(r_1, \| u_n(t') - x_1 \|) < \epsilon$. Then, by the triangle inequality: $\| u_n(t) - u^-(t) \| < 3\epsilon$ and, in the limit, $\| u(t) - u^-(t) \| \leq 3\epsilon$. Thus, $u^-(t) = u(t)$.

3)   To end the proof that u is the unique solution to the sweeping process by C starting at a (and hence that the whole sequence (u_n) has a unique sublimit and therefore converges pointwisely), we can now apply Proposition 2.2. We only need to establish (6).

Let $\phi : [0, T] \to \mathbb{R}^d$ be a continuous selection of C and let $s < t$. From Proposition 2.5, we may take a sequence (ϕ_n) converging uniformly to ϕ, with ϕ_n a continuous selection of C_n . Since ϕ_n is also a selection of C_m for $m \geq n$, then, by the necessary part of Proposition 2.2, u_m satisfies:

$$\int_{]s,\,t]} \phi_n \cdot du_m \;\geq\; \tfrac{1}{2}\big(\,\|\,u_m(t)\,\|^{\,2} - \|\,u_m(s)\,\|^{\,2}\big);$$

taking limits as $m \to \infty$ (use Theorem 0.2.1), we get:

$$\int_{]s,\,t]} \phi_n \cdot du \;\geq\; \tfrac{1}{2}\big(\,\|\,u(t)\,\|^{\,2} - \|\,u(s)\,\|^{\,2}\big);$$

and since $\phi_n \to \phi$ uniformly:

$$\int_{]s,\,t]} \phi \cdot du \;\geq\; \tfrac{1}{2}\big(\,\|\,u(t)\,\|^{\,2} - \|\,u(s)\,\|^{\,2}\big). \qquad\qquad \square$$

Notice that if the graph of C is not left-closed, then the solution to the sweeping process by C may not be continuous. At discontinuity points, the converse statement in Proposition 2.2 does not hold and the convergence result of Theorem 2.6 ceases to be valid, as shown by Valadier:

Example 2.7 [Val 1,2] Let $\Gamma(\alpha) = \mathrm{co}\,\{(-1,\alpha),(0,0),(0,-1),(-1,-1)\}$, if $\alpha \in [0,1]$; and for $\alpha \geq 1$ let $\Gamma(\alpha) = (-\tfrac{1}{2},-\tfrac{1}{2}) + \alpha\,[\Gamma(1) + (\tfrac{1}{2},\tfrac{1}{2})]$ — in other words, $\Gamma(\alpha)$ is the transform of $\Gamma(1)$ in the similarity of center $(-\tfrac{1}{2},-\tfrac{1}{2})$ and ratio α (a picture describing Γ is easily drawn). Since Γ is Lipschitz-continuous and increasing, then

$$(10) \qquad\qquad C_n(t) := \Gamma(\,n(1-t)\,) \quad (t \in [0,1])$$

is also Lipschitz-continuous ($\forall n$) and increasing: $C_n(t) \subset C_{n+1}(t)$. It is clear that, in the sense of Proposition 2.5, (C_n) converges to the multifunction defined by:

$$(11) \qquad\qquad C(t) = \begin{cases} \mathbb{R}^2 & \text{if } 0 \leq t < 1 \\ \Gamma(0) = [-1,0] \times [-1,0] & \text{if } t = 1 \end{cases}.$$

Take $a = (0,\sqrt{2})$ as the initial condition. For n large enough, $a \in C_n(0) = \Gamma(n)$ and we may consider the solution u_n to the sweeping process by C_n starting from a. This solution u_n presents an initial immobility at a, with growing duration (it lasts until the ratio of similarity $\alpha = n(1-t)$ equals $(1+\tfrac{\sqrt{2}}{2})(\tfrac{\sqrt{2}}{2})^{-1}$, i.e. for $t \leq 1 - \tfrac{1+\sqrt{2}}{n}$). After that, u_n moves on a straight line (normal to the boundary of the moving convex set) down to $(-\tfrac{\sqrt{2}}{2},\tfrac{\sqrt{2}}{2})$, attained at time $t = 1 - \tfrac{1}{n}$, and then, pushed by the boundary of $\Gamma(\alpha)$ ($\alpha \to 0$) it reaches $(-1,0)$ at $t=1$ by a circular motion. Thus, the pointwise limit of (u_n) is

$$\overline{u}(t) = \begin{cases} (0,\sqrt{2}) & \text{if } 0 \leq t < 1 \\ (-1,0) & \text{if } t = 1 \end{cases},$$

while the solution to the sweeping process by C is:

$$u(t) = \begin{cases} (0, \sqrt{2}) & \text{if } 0 \leq t < 1 \\ (0, 0) & \text{if } t = 1 \end{cases},$$

since the projection of $(0, \sqrt{2})$ in $C(1)$ is $(0, 0)$. $\square$

We now turn our attention to **Lipschitz approximations from the exterior**, following Gavioli's work [Gav].

We say that a multifunction C is *metrically upper semicontinuous* at t if $\tau \to t$ implies that $e(C(\tau), C(t)) \to 0$, or equivalently, if for every positive ϵ, in some neighbourhood $U = U(\epsilon)$ of t:

$$(12) \qquad\qquad C(\tau) \subset C(t) + \epsilon \overline{B}(0, 1), \quad \forall \tau \varepsilon\, U.$$

Theorem 2.8 [Gav] (Theorem 2.1) *Let C be a metrically upper semicontinuous multifunction from a metric space (I, d) to nonempty bounded closed convex subsets of a normed space E. Assume that C is bounded in the sense that* $C(t) \subset \overline{B}(0, M)$, $\forall t\varepsilon\, I$.
Then, there exist Lipschitz-continuous multifunctions C_n $(n \geq 1)$, of the same type, with the properties:

$$(13) \qquad\qquad C_n(t) \supset C_{n+1}(t) \supset C(t) , \quad \forall t\varepsilon\, I ;$$

$$(14) \qquad\qquad h(C_n(t), C(t)) \to 0 , \quad \forall t\varepsilon\, I ;$$

$$(15) \qquad\qquad h(C_n(t), C_n(s)) \leq L_n\, d(t, s) , \quad \forall t, s\varepsilon\, I .$$

The technique is essentially geometrical and the following result is needed. For simplicity, we write $\overline{B} = \overline{B}(0, 1)$.

Lemma 2.9 [Gav] (Lemma 2.1) *Let $(\Gamma_i)_{i \varepsilon I}$ be a family of nonempty bounded closed convex subsets of E. Assume that all the sets Γ_i contain a fixed ball $\overline{B}(x_0, h)$ and write*

$$(16) \qquad\qquad L = \inf_i\, |\Gamma_i - x_0| = \inf_i\, \sup_{x \varepsilon \Gamma_i} \| x - x_0 \| .$$

Then:

$$(17) \qquad\qquad \forall\, \epsilon > 0, \quad \bigcap_{i \varepsilon I} (\Gamma_i + \epsilon \overline{B}) \subset \bigcap_{i \varepsilon I} \Gamma_i + \frac{L}{h} \epsilon \overline{B} ;$$

i.e., the intersection of enlarged sets is contained in an enlarged intersection of the sets Γ_i .

Proof. Writing $\Gamma = \bigcap_i \Gamma_i$ and $\Gamma_\epsilon = \bigcap_i (\Gamma_i + \epsilon \overline{B})$, we show that, for every $x \epsilon \Gamma_\epsilon$, there is some $x' \epsilon \Gamma$ with $\| x' - x \| \leq L \frac{\epsilon}{h}$, by taking

$$x' := x + \frac{\epsilon}{h+\epsilon} (x_0 - x) .$$

In fact, $x \epsilon \Gamma_\epsilon$ implies that $\| x - x_0 \| \leq | \Gamma_i - x_0 | + \epsilon$ for every i, hence $\| x' - x \| \leq \epsilon (h+\epsilon)^{-1} (L+\epsilon) \leq \epsilon L/h$ (since $L \geq h$). On the other hand, for every i, there is some $x_i \in \Gamma_i \cap (x + \epsilon \overline{B})$. Then the point $z_i := x' - \frac{h}{\epsilon} (x_i - x')$ satisfies $\| z_i - x_0 \| = \| \frac{h}{\epsilon} (x - x_i) \| \leq h$ and so $z_i \epsilon \Gamma_i$. Hence, remarking that x' is a convex combination of x_i and z_i, to be precise:

$$x' = \frac{\epsilon}{h+\epsilon} z_i + \frac{h}{h+\epsilon} x_i ,$$

we conclude that $x' \epsilon \Gamma_i$ (for every i) , i.e., $x' \epsilon \Gamma$. $\square$

The following Lemma contains some properties needed in the sequel. It is a particularization of a standard regularization or approximation method for real functions on metric spaces.

Lemma 2.10. *Define functions* r *and* r_n *by* $r(t, \tau) := e(C(t) , C(\tau))$ *and* $r_n(t, \tau) = \frac{1}{n} + \sup_{t' \epsilon I} [r(t', \tau) - n\, d(t, t')]$ *. Then they satisfy*

(18) $\rho > r(t, \tau) \Rightarrow C(t) \subset C(\tau) + \rho \overline{B}$;

(19) $r_n(t, \tau) > r_{n+1}(t, \tau) , \quad 1+2M \geq r_n(t, \tau) \geq \frac{1}{n} + r(t, \tau)$;

(20) $r_n(t, \tau) \to e(C(t), C(\tau))$;

(21) $| r_n(t, \tau) - r_n(s, \tau) | \leq n\, d(t, s)$.

Proof. The definition of $r(t, \tau)$ as the excess of $C(t)$ over $C(\tau)$ gives (18). It is obvious that (r_n) is a decreasing sequence. Taking $t' = t$, we obtain $r_n(t, \tau) \geq \frac{1}{n} + r(t, \tau)$. Moreover, $C(t') \cup C(\tau) \subset \overline{B}(0, M)$ implies $r(t', \tau) \leq 2M$ and so $r_n \leq 1 + 2M$.

To prove (20), take $\epsilon > 0$ and a positive number δ such that $| r(t', \tau) - r(t, \tau) | < \epsilon$ whenever $d(t', t) < \delta$. If $n \geq 2M/\delta$, then $d(t', t) \geq \delta$ implies $r(t', \tau) - n\, d(t, t') \leq 2M - n\delta \leq 0$; so, these t' do not matter in the computation of r_n . It follows that

$$\tfrac{1}{n} + r(t, \tau) \leq r_n(t, \tau) \leq \tfrac{1}{n} + r(t, \tau) + \epsilon$$

and eventually $| r_n(t, \tau) - r(t, \tau) | \leq 2\epsilon$, proving convergence.

Writing $\overline{r}_n = r_n - \frac{1}{n}$, we have, for every $t' \varepsilon\, I$:

$$\overline{r}_n(t,\tau) \geq r(t',\tau) - n\, d(t',t) \geq r(t',\tau) - n\, d(t',s) - n\, d(s,t) \, ,$$

so that, by taking the supremum

$$\overline{r}_n(t,\tau) \geq \overline{r}_n(s,\tau) - n\, d(s,t);$$

exchanging t and s, we obtain $\;| \overline{r}_n(t,\tau) - \overline{r}_n(s,\tau) | \leq n\, d(s,t)\;$ and (21) follows. $\square$

We can now proceed with the proof of the approximation theorem.

Proof of Theorem 2.8. Define for every t in the metric space (I, d):

$$(22) \qquad\qquad C_n(t) = \mathrm{cl} \bigcap_{\tau\,\varepsilon\, I} (C(\tau) + r_n(t,\tau)\,\overline{B}) \, ,$$

a bounded closed convex subset of E.

By the preceding Lemma, (18) and (19), $r_n(t,\tau) > r(t,\tau)$ implies $C_n(t) \supset C(t)$. Because r_n is nonincreasing, the sequence $(C_n(t))$ is also nonincreasing, thus proving (13).

To prove (14), we fix $t\varepsilon\, I$, $\epsilon > 0$ and, by (20), we choose $\overline{n}$ so that $r_n(t,t) < \epsilon$ for $n \geq \overline{n}$. Then $C(t) \subset C_n(t) \subset \mathrm{cl}\,(C(t) + r_n(t,t)\,\overline{B}) \subset C(t) + \epsilon\,\overline{B}$ yields $h(C_n(t), C(t)) \leq \epsilon$.

In order to prove that C_n is a Lipschitz-continuous multifunction (claim (15)) we write

$$\Gamma_n(t,\tau) = C(\tau) + r_n(t,\tau)\,\overline{B} \; ;$$

from (21) it follows that:

$$(23) \qquad\qquad \Gamma_n(s,\tau) \subset \Gamma_n(t,\tau) + n\, d(t,s)\,\overline{B}.$$

The family $\{\Gamma_n(t,\tau) : \tau\varepsilon\, I\}$ fulfils the assumption of Lemma 2.9. Indeed, for any $x_0\,\varepsilon\, C(t)$, by (18) we have

$$\overline{B}(x_0, \tfrac{1}{2n}) \subset C(t) + \tfrac{1}{2n}\overline{B} \subset C(\tau) + (r(t,\tau) + \tfrac{1}{2n})\,\overline{B} + \tfrac{1}{2n}\overline{B}$$

and so, by (19):

$$(24) \qquad\qquad \overline{B}(x_0, \tfrac{1}{2n}) \subset C(\tau) + r_n(t,\tau)\,\overline{B} = \Gamma_n(t,\tau);$$

moreover, $\Gamma_n(t,t) - x_0 \subset C(t) + r_n(t,t)\,\overline{B} - C(t) \subset [\,M + (2M+1) + M\,]\,\overline{B}$ implies that:

$$L= \inf_{\tau} \ |\Gamma_n(t,\tau)-x_0| \ \leq \ |\Gamma_n(t,t)-x_0| \ \leq \ 4\,M+1.$$

Hence, by (23) and applying Lemma 2.9 with $h=\frac{1}{2n}$ and $\epsilon=n\,d(t,s)$:

$$\bigcap_{\tau} \Gamma_n(s,\tau) \subset \bigcap_{\tau} (\Gamma_n(t,\tau)+\epsilon\,\overline{B})$$

$$\subset (\bigcap_{\tau}\Gamma_n(t,\tau))+2\,n\,L\,\epsilon\,\overline{B} \subset C_n(t)+2\,n^2\,(4M+1)\,d(t,s)\,\overline{B}.$$

Thus $C_n(s) \subset C_n(t)+L_n\,d(t,s)\,\overline{B}$ and a similar inclusion holds when s and t are exchanged. Thus, $h(\,C_n(s),\,C_n(t)\,) \leq \ L_n\,d(t,s)$, with $L_n=2\,n^2\,(4M+1)\,.$ $\quad\Box$

According to [Gav], Theorem 2.8 is still true for multifunctions with "sublinear growth", that is, with $|\,C(t)\,| \ \leq \ M(\,1+d(t,t_0)\,)\,.$

A metrically upper semicontinuous multifunction C can thus be approached by a sequence (C_n) of Lipschitz-continuous multifunctions for which there exist solutions (u_n) to the sweeping process (Theorem 2.1). If additionally the multifunction C is Hausdorff-continuous and if $C(t)$ has nonempty interior (for every t) then it can be shown that (u_n) converges uniformly to the solution to the sweeping process by C:

Theorem 2.11 [Gav] (Theorem 3.1) *Let C be a multifunction on $I=[0,T]$ with bounded closed convex values in a separable Hilbert space H. Assume that C is Hausdorff-continuous (i. e., $s \to t \Rightarrow h(\,C(s),\,C(t)) \to 0$) and that every $C(t)$ has nonempty interior. Take a sequence (C_n) of Lipschitz-continuous multifunctions approaching C from the exterior as in Theorem 2.8 and, $a \ \varepsilon \ C(0)$ being fixed, denote by u_n the solution to the sweeping process by C_n with $u_n(0)=a$.*

Then, (u_n) converges uniformly on I to a continuous bv function u, which is the solution to the sweeping process by C with initial value $u(0)=a$.

Proof. As in the proof of Lemma 2.3.2(b), we may divide I into a finite number of intervals $J_k=[t_k,t_{k+1}]$ such that all the sets $C(t)$ with t in J_k contain some fixed ball $\overline{B}(a_k,r_k)$. Since the same can be said of $C_n(t)$ – which contains $C(t)$ – for $n\varepsilon\mathbb{N}$ and $t\varepsilon\,J_k$, then, from Theorem 2.1, (5), we get $\mathrm{var}\,(u_n,J_k) \leq \ (2\,r_k)^{-1}\,\|\,u_n(t_k)-a_k\,\|^2$. Therefore, it is easily shown by induction that (u_n) is uniformly bounded both in norm and in variation, say by $M>0\,.$

We now prove that these Lipschitz-continuous functions converge uniformly to some function u, which must be continuous and have $\mathrm{var}\,u \leq M$.

Let $m < n$. Since $u_m(0) = u_n(0) = a$, we have

$$\tfrac{1}{2} \| u_n(t) - u_m(t) \|^2 = \int_0^t (u_n(s) - u_m(s)) \cdot (u_n'(s) - u_m'(s)) \, ds$$

and, by taking $u_{m,n}(s)$ to be the projection of $u_m(s)$ in $C_n(s)$, we rewrite the above integral as the following sum:

$$(25) \quad \int_0^t (u_n(s) - u_{m,n}(s)) \cdot u_n'(s) \, ds + \int_0^t (u_{m,n}(s) - u_m(s)) \cdot u_n'(s) \, ds$$

$$+ \int_0^t (u_m(s) - u_n(s)) \cdot u_m'(s) \, ds.$$

The first term is not positive, since u_n solves the sweeping process by C_n and $u_{m,n}$ is a (continuous) selection of C_n (cf. Proposition 2.2). Since $u_n(s) \,\varepsilon\, C_n(s) \subset C_m(s)$ and u_m is a solution to the sweeping process by C_m, the third integral in (25) is also nonpositive. Notice that:

$$\| u_{m,n}(s) - u_m(s) \| = \mathrm{dist}\,(u_m(s), C_n(s))$$

$$\leq \mathrm{dist}\,(u_m(s), C(s)) \leq \phi_m(s) := h(C_m(s), C(s)),$$

since $C_n(s) \supset C(s)$ and $u_m(s) \,\varepsilon\, C_m(s)$. Thus, we have

$$(26) \quad \tfrac{1}{2} \| u_n(t) - u_m(t) \|^2 \leq \int_0^t (u_{m,n}(s) - u_m(s)) \cdot u_n'(s) \, ds$$

$$\leq \| \phi_m \|_\infty \, \mathrm{var}\,(u_n) \leq M \, \| \phi_m \|_\infty \, .$$

The ϕ_m are continuous functions (C_n and C are continuous multifunctions), converging monotonically to zero, as $m \to \infty$ (under the conditions of Theorem 2.8, (13), (14)). By Dini's theorem, $\| \phi_m \|_\infty \to 0$. Thus, (26) implies that (u_n) is indeed a uniformly Cauchy hence convergent sequence.

The limit function u satisfies the initial condition and $u(t) \,\varepsilon\, C(t)$, since

$$\mathrm{dist}\,(u(t), C(t)) = \lim_n \mathrm{dist}\,(u_n(t), C(t)) \leq \lim_n h(C_n(t), C(t)) = 0$$

and $C(t)$ is closed.

In view of Proposition 2.2, to complete the proof that the cbv function u is the solution to the sweeping process by C, we only need to show that (6) holds for every continuous selection of C, say ϕ. In fact, any such ϕ is also a continuous selection of C_n, so:

$$\int_{]s,t]} \phi \cdot du_n \geq \tfrac{1}{2} \| u_n(t) \|^2 - \tfrac{1}{2} \| u_n(s) \|^2 \, .$$

Passing to the limit (we use uniform convergence or Theorem 0.2.1), we get

$$\int_{]s,\,t]} \phi \cdot du \;\geq\; \tfrac{1}{2}\,\|\,u(t)\,\|^2 - \tfrac{1}{2}\,\|\,u(s)\,\|^2\,,$$

as required. $\square$

5.3. An application of differential inclusions to quasi-statics

Following Chraibi [Chr], we study the motion, with respect to an orthonormal set of coordinates $(O,\,\overrightarrow{e}_1,\,\overrightarrow{e}_2,\,\overrightarrow{e}_3)$, of a point $q(t)$ $(t\varepsilon I:=[0,\,T])$ which is drawn to a point with known motion $a(t)$ by an elastic force $F=-\,k\,(q(t)-a(t))$ where $k>0$ is fixed. The point q is to be confined to the half-space

$$L=\{\,(x_1,\,x_2,\,x_3)\,\varepsilon\,\mathbb{R}^3:\ x_3\leq 0\,\}$$

by a fixed rigid wall P with equation $x_3=0$. When q comes into contact with P, we assume that it is subjected to dry friction of Coulomb type, isotropic but possibly non-homogeneous: the friction coefficient ν may depend on the position q.

Moreover, we place ourselves in a quasi-statical setting: the mass of q is taken to be zero. This simplifying assumption is reasonable in many Engineering applications and it leads to a reduced size of computations. However, the smoothing effect of inertia is lost and so the study of the corresponding mathematical problems may need some Nonsmooth Analysis.

We prove the existence of a solution by showing that the problem is equivalent to solving a sweeping process by a multifunction which depends on the actual position of the moving point; thus, we have to find a fixed point. Recall that a similar fixed point scheme was (unsuccessfully) proposed in Chapter 4.

First, we give an appropriate formulation of the problem.

The regularity assumptions are the following: the motion of the center of attraction $a(\,.\,)$ is absolutely continuous and the friction coefficient $\nu:P\to\,]0,+\infty[$ is a bounded and Lipschitz-continuous function of q: $\nu(q)\leq m,\ |\,\nu(q)-\nu(q')\,|\leq k'\,\|\,q-q'\|$ (we may suppose that ν is extended to the whole space). Let $q_0\,\varepsilon\,L$ be given.

Definition 3.1. We say that an absolutely continuous function $q: I \to \mathbb{R}^3$ is a **solution to Problem QS** if:

$$(1) \qquad\qquad q(0) = q_0 \, ,$$

and, for some choice of the representative of the derivative $\dot{q}$:

$$(2) \qquad\qquad -\dot{q}(t) \, \varepsilon \, N_{\mathcal{C}_q(t)}(q(t)) \qquad (\forall t \varepsilon I)$$

where $\mathcal{C}_q : I \to \mathbb{R}^3$ is the multifunction defined by

$$(3) \qquad \mathcal{C}_q(t) = \left\{ \begin{array}{ll} \{\, a(t) \,\} & \text{if } a_3(t) \le 0 \\ D_q(t) = D(\, b(t),\, a_3(t)\,\nu(q(t))\,) & \text{if } a_3(t) > 0 \end{array} \right. .$$

Here, $D_q(t)$ denotes the disk contained in P with center $b(t) = (a_1(t), a_2(t), 0)$ and radius $a_3(t)\,\nu(q(t))$.

Remark 3.2. We show that this formulation is compatible with the usual one. We introduce the *reaction force R* which must equilibrate the force F; i.e., $F + R = 0$ or

$$(4) \qquad\qquad R = k(q - a) \, .$$

Notice that (2) implicitly requires that

$$(5) \qquad\qquad q(t) \, \varepsilon \, \mathcal{C}_q(t) \subset L \quad (\forall \, t \varepsilon I) \, .$$

1) In particular, by the definition of $\mathcal{C}_q$, if $q(t)$ belongs to the interior of the admissible region L (i.e., $q_3(t) < 0$) then $a_3(t) < 0$ and moreover $q(t) = a(t)$. That is,

$$(6) \qquad\qquad q_3 < 0 \;\Rightarrow\; R = 0 \;\;(\,\text{and } q = a\,) \, ,$$

as prescribed by the friction law.

2) If $q(t) \varepsilon P$ and there are $t_n \to t$ with $q(t_n) \notin P$, then, since $a(t_n) = q(t_n)$, we again conclude that $q(t) = a(t)$ and $R(t) = 0$. This is compatible with the friction law, since the reaction belongs to the **friction cone** at $q = q(t)$:

$$(7) \qquad C(q) := \{ (r_1, r_2, r_3) : \; r_3 \le 0 \,, \; \| (r_1, r_2) \| \le -\nu(q)\,r_3 \} \, .$$

3) Finally, let $q(s) \varepsilon P$ in some neighbourhood of t (in $[0, T]$). We must prove that, in the syntethic formulation of the friction law by Moreau (compare with Chapter 3), for almost every such t:

$$(8) \qquad\qquad -\dot{q}(t) \, \varepsilon \, \text{proj}_P \, N_{C(q(t))}(R(t)) \, ,$$

the orthogonal projection into P of the outward normal cone to the friction cone $C(q(t))$ at the reaction $R(t)$. First, it is clear that

$$\dot{q}(t) \, \varepsilon \, P, \tag{9}$$

if the derivative exists, which happens for almost every t. By (5), we have $a_3(t) \geq 0$. Moreover, if $a_3(t) = 0$, then $R(t) = 0$ as above and it is easily seen that the outward normal cone to $C(q(t))$ at $R = 0$ is a nontrivial cone (contained in the half-space $-L$) which projects itself **onto** P. Thus (8) is obviously satisfied.

Assume that $a_3(t) > 0$, hence $a_3(s) > 0$ in a neighbourhood of t. By (5), $q(s) \, \varepsilon \, C_q(s) = D_q(s)$. Two situations may occur.

Stick — If $\| q(t) - b(t) \| < a_3(t) \, \nu(q(t))$, i.e., if $q(t) \, \varepsilon \, \mathrm{relint} \, C_q(t)$, then (2) and (9) imply that $\dot{q}(t) = 0$. On the other hand, the reaction $R(t) = k(q(t) - a(t))$ is in the interior of the friction cone: in fact, $R_3(t) = -k \, a_3(t) < 0$ and $\| (R_1(t), R_2(t)) \| = k \| q(t) - b(t) \| < k \, a_3(t) \, \nu(q(t)) = -\nu(q(t)) \, R_3(t)$. Thus, the normal cone to $C(q(t))$ at $R(t)$ is zero and (8) is satisfied.

Slip — Let $\| q(t) - b(t) \| = a_3(t) \, \nu(q(t)) > 0$. The computations above show that the nonzero reaction $R(t)$ belongs to the boundary of the friction cone. The projection into P of the outward normal cone $N_{C(q(t))}(R(t))$ is then easily found to be the half-line $\{ \lambda (R_1(t), R_2(t), 0) \; : \; \lambda \geq 0 \}$. On the other hand, $N_{C_q(t)}(q(t)) \cap P$ is the half-line spanned by $q(t) - b(t)$, so that (2) and (9) give

$$-\dot{q}(t) = \mu \, (q(t) - b(t)) \qquad (\mu > 0) \, .$$

Thus $-\dot{q}(t) = \mu \, (q_1(t) - a_1(t), \, q_2(t) - a_2(t), \, 0) = \frac{\mu}{k} \, (R_1(t), \, R_2(t), \, 0)$ does indeed satisfy (8). $\qquad\qquad \square$

The formulation of Problem QS is an invitation to a fixed point scheme and actually this is how we prove the existence theorem that follows ([Chr], Ch. IV, although not so clearly stated). We say that the initial position q_0 is *equilibrated* if it belongs to $C_{q_0}(0)$, that is, $q_0 = a(0)$ if $a_3(0) \leq 0$ or $q_0 \, \varepsilon \, D(b(0), a_3(0) \nu(q_0))$ if $a_3(0) > 0$. Recall that k' is the Lipschitz constant of $\nu(.)$ and that $a(.)$ is an absolutely continuous function.

Theorem 3.3. [Chr] *If q_0 is equilibrated and $\kappa := k' \| a_3 \|_\infty < 1$, then Problem QS has an absolutely continuous solution defined on $[0, T]$.*

Proof. Consider the set K of continuous functions r from $[0, T]$ to $\mathbb{R}^3$ such that $r(0) = q_0$ and

$$(10) \qquad \| r(t) - r(s) \| \leq \frac{A}{1-\kappa} \, \mathrm{var}\,(a; s, t) \qquad (\forall \, 0 \leq s \leq t \leq T) \,,$$

where A is the constant given by Lemma 3.10 below. It is obvious that K is a nonempty convex set, which is closed for uniform convergence; moreover, it is equibounded (since $\| r(t) \| \leq \| q_0 \| + A\,(1-\kappa)^{-1} \, \mathrm{var}\,(a; 0, T)$) and by (10) it is equicontinuous, since $a(\,.\,)$ is absolutely continuous; thus, by Ascoli-Arzelà's theorem, K is a compact subset of $\mathcal{C}(I, \mathbb{R}^3)$.

Note that if $r \varepsilon K$ then r is absolutely continuous and by Lemma 3.10 below, $t \to \mathcal{C}_r(t)$ (defined as in (3)) is an absolutely continuous multifunction. To be precise,

$$(11) \quad \mathrm{var}\,(\mathcal{C}_r;\, s, t) \leq A\, \mathrm{var}(a;\, s,\, t) + \kappa\, \frac{A}{1-\kappa} \, \mathrm{var}\,(a;\, s,\, t) = \frac{A}{1-\kappa} \, \mathrm{var}\,(a;\, s,\, t) \,.$$

Hence, by Moreau's results on sweeping processes (see e.g. [Mor 1]), to every $r \varepsilon K$ we may associate $\phi(r) = q_r$, the unique absolutely continuous solution to the sweeping process by $\mathcal{C}_r(t)$:

$$q_r(0) = q_0 \, , \; -\dot{q}_r(t) \,\varepsilon\, N_{\mathcal{C}_{r(t)}}(q_r(t)) \;\; \text{a.e. } t \,.$$

In order to apply Leray-Schauder's fixed point theorem, we have to prove that $\phi(r) \varepsilon K$ and that $r \to \phi(r)$ is continuous in K. In fact, $\phi(r)$ is a continuous function, $\phi(r)(0) = q_0$ and, by [Mor 1], §2.c (the variation of the solution is not greater than that of the multifunction) and by (11), the following estimate holds:

$$\| \phi(r)(s) - \phi(r)(t) \| \; \leq \; \mathrm{var}\,(\mathcal{C}_r;\, s, t) \; \leq \; \frac{A}{1-\kappa} \, \mathrm{var}\,(a;\, s,\, t) \,,$$

showing that $\phi(r) \varepsilon K$.

To prove that ϕ is continuous, take a sequence $(r_n) \subset K$ converging uniformly to $r \varepsilon K$. By the results of Moreau on the dependence of solutions to sweeping processes on the data (see [Mor 1], Prop. 2.g, (2.16)), we have:

$$\| \phi(r_n)(t) - \phi(r)(t) \|^2 \; \leq \; 2\,\mu(t)\,[\mathrm{var}\,(\mathcal{C}_{r_n};\, 0, t) + \mathrm{var}\,(\mathcal{C}_r;\, 0, t)] \,,$$

where $\mu(t)$ is the least upper bound of the Hausdorff distance $h(\mathcal{C}_{r_n}(s), \mathcal{C}_r(s))$ for $s \varepsilon [0, t]$. By Lemma 3.11, $\mu(t) \leq \kappa \max\{\, \| r_n(s) - r(s) \| :\; 0 \leq s \leq t\}$ $\leq \kappa \, \| r_n - r \|_\infty$, while (11) gives estimates for the variations appearing above. It follows that

$$\| \phi(r_n) - \phi(r) \|_\infty^2 \leq \frac{4\kappa A}{1-\kappa} \, \mathrm{var}\,(a;\, 0,\, T) \, \| r_n - r \|_\infty \,.$$

Hence $r_n \to r$, uniformly, implies that $\phi(r_n) \to \phi(r)$ uniformly on $[0,\, T]$.

By Leray-Schauder's theorem, there is a function $q \varepsilon K$ such that $\phi(q) = q$. Clearly, q is a solution to Problem QS. $\qquad\qquad\square$

Remark 3.4. Notice that the proof yields the estimate

$$(12) \qquad\qquad \| q(t) - q(s) \| \leq \frac{A}{1-\kappa} \, \mathrm{var}\,(a;\, s,\, t) \,.$$

Remark 3.5. In a 2-dimensional setting, Chraibi ([Chr], Chapter II, §2) exhibited numerical examples showing that if $\kappa > 1$ then discontinuities of the motion may appear. The condition $\kappa < 1$ — which, roughly put, says that the normal distance of the attracting point and the variation of the friction coefficient cannot be simultaneously large — has thus a mechanical (or at least a numerical) meaning.

Remark 3.6. If the friction coefficient ν is independent of q ("homogeneous friction"), then Problem QS becomes a simple sweeping process:

$$(13) \qquad\qquad -\dot{q}(t) \,\varepsilon\, N_{C(t)}(q(t)) \,,$$

where

$$(14) \qquad\qquad C(t) := \begin{cases} \{\, a(t)\,\} & \text{if } a_3(t) \leq 0 \\ D(\,b(t),\, \nu\, a_3(t)\,) & \text{if } a_3(t) > 0 \end{cases}$$

is an absolutely continuous multifunction, as in the proof of the theorem. Hence ([Chr], Chapter I) there is a unique solution and it can be approached by a discretization procedure:

$$(15) \qquad\qquad q_n(0) = q_0 \,, \quad q_n(t_{n,i+1}) = \mathrm{proj}\,(q_n(t_{n,i}),\, C(t_{n,i+1})) \,.$$

Remark 3.7. In [Chr], Chapter II, a two-dimensional version of the existence theorem is given. In that case, even considering different "right" and "left" friction coefficients, the simpler Banach fixed point theorem for contractions is enough to yield existence of a unique solution.

Remark 3.8. In [Chr], Chapter III, a viscosity $v > 0$ is introduced in the equation:

$$(16) \qquad\qquad v\, \dot{q} + k(q - a) = R \,.$$

Assuming a persistent contact, the problem is equivalent to a differential equation:

$$(17) \qquad \dot{q}(t) = \text{proj}\Big(0,\ D\big(\tfrac{k}{v}(b(t) - q(t)),\ \tfrac{k}{v}\,a_3(t)\,\nu(q(t))\big)\Big)$$

and existence follows by the Cauchy-Peano theorem. In this context, the following inequality about projections of a point into two different plane disks is useful (see [Chr], Annexe III):

$$(18) \qquad \|\,\text{proj}\,(q, D_1) - \text{proj}\,(q, D_2)\,\| \ \leq\ h(D_1, D_2)\,.$$

Remark 3.9. If we assume persistent contact (for instance, by taking $q_0 \varepsilon\, P$ and $a_3(t) > 0$, for all t) then we only need to consider the case of the disk in the definition of $C_q(t)$. Therefore, if the attracting point $a(\,.\,)$ is merely continuous we may apply the results of Chapter 2 on sweeping processes by continuous convex sets with nonempty interior. These immediately yield existence for the case of an homogeneous friction coefficient: the solution is (at least) continuous with bounded variation.

A further investigation of these quasi-static problems is currently under way by J. Martins, F. Gastaldi and the author (see [Mar & al] and [Mon 9]).

We end this section with two technical lemmas used in the preceding proofs.

Lemma 3.10. [Chr] (Annexe IV) *There is a positive number A, depending only on the friction coefficient ν, such that*

$$(19) \qquad h(C_r(t), C_r(s)) \ \leq\ A\,\|\,a(t) - a(s)\,\| + \kappa\,\|\,r(t) - r(s)\,\|\,,$$

for every function r from $I = [0, T]$ to $\mathbb{R}^3$ and every $s, t \varepsilon\, I$.
Hence, the variation of $t \to C_r(t)$, in the sense of Hausdorff distance, satisfies

$$(20) \qquad \text{var}\,(C_r\,;\, s, t) \ \leq\ A\,\text{var}\,(a\,;\, s, t) + \kappa\,\text{var}\,(\,r\,;\, s, t)\,.$$

Proof. 1) If both $a(t)$ and $a(s)$ belong to the admissible region L, then $h(C_r(t), C_r(s)) = \|\,a(t) - a(s)\,\|$ and (19) holds with any $A \geq 1$.

2) If $a(t) \notin L$ and $a(s) \varepsilon\, L$, then the Hausdorff distance between $C_r(t) = D(b(t), a_3(t)\,\nu(r(t)))$ and $C_r(s) = \{\,a(s)\,\}$ is the greatest distance between $a(s)$ and a point of the disk. Thus,

(21)
$$h(\mathcal{C}_r(t), \mathcal{C}_r(s)) \leq \|\, a(s) - b(t)\,\| + a_3(t)\,\nu(r(t)).$$

Notice that $(a(s) - b(t)) \cdot (a(t) - b(t)) = (a(s) - b(t)) \cdot (0, 0, a_3(t)) = a_3(s)\,a_3(t) \leq 0$; therefore, the angle between those two vectors is $\geq \frac{\pi}{2}$ and the opposite side of the triangle (defined by $a(s)$, $b(t)$ and $a(t)$) has the greatest length:

(22)
$$\|\, a(s) - b(t)\,\| \leq \|\, a(s) - a(t)\,\|.$$

Moreover, $a_3(s) \leq 0$ and $0 < \nu(r(s)) \leq m$ imply that:

$$a_3(t)\,\nu(r(t)) \leq a_3(t)\,\nu(r(t)) - a_3(s)\,\nu(r(s))$$

$$\leq a_3(t)\,|\,\nu(r(t)) - \nu(r(s))\,| + |\,a_3(t) - a_3(s)\,|\,\nu(r(s))$$

$$\leq \|\, a_3\,\|_\infty\, k'\, \|\, r(t) - r(s)\,\| + m\, \|\, a(t) - a(s)\,\|.$$

Combining with (21) and (22) and recalling that $\kappa = \|\, a_3\,\|_\infty\, k'$, we see that (19) holds with $A = 1 + m$.

3) Assume that $a_3(t)$ and $a_3(s)$ are both positive. Then we have to consider the Hausdorff distance h between disks with centers $b(t) = (a_1(t), a_2(t), 0)$ and $b(s) = (a_1(s), a_2(s), 0)$ and radii $\rho(t) = a_3(t)\,\nu(r(t))$ and $\rho(s) = a_3(s)\,\nu(r(s))$. It is known (and easily seen) that

$$h = \|\, b(t) - b(s)\,\| + |\,\rho(t) - \rho(s)\,|\;;$$

since $\|\, b(t) - b(s)\,\| \leq \|\, a(t) - a(s)\,\|$ (from the definition of $b(.)$) and since $|\,\rho(t) - \rho(s)\,|$ can be estimated as in the second case, it follows that:

$$h \leq \|\, a(t) - a(s)\,\| + \kappa\, \|\, r(t) - r(s)\,\| + m\, \|\, a(t) - a(s)\,\|,$$

and again $A = 1 + m$ will do the job. $\qquad\square$

Lemma 3.11. *Given two functions r and q, from $I = [0, T]$ to $\mathbb{R}^3$, we have:*

(23)
$$h(\mathcal{C}_r(t), \mathcal{C}_q(t)) \leq \kappa\, \|\, r(t) - q(t)\,\| \qquad (\forall t \varepsilon I).$$

Proof. If $a_3(t) \leq 0$, then $\mathcal{C}_r(t) = \mathcal{C}_q(t) = \{\, a(t)\,\}$ and the result is obvious.

If $a_3(t) > 0$, then taking $\rho(t) = a_3(t)\,\nu(r(t))$ and $\overline{\rho}(t) = a_3(t)\,\nu(q(t))$ we have:

$$h(\mathcal{C}_r(t), \mathcal{C}_q(t)) = h(\, D(b(t), \rho(t)),\, D(b(t), \overline{\rho}(t))\,) = |\,\rho(t) - \overline{\rho}(t)\,|$$

$$\leq \|\, a_3\,\|_\infty\, k'\, \|\, r(t) - q(t)\,\|. \qquad\square$$

5.4 Additional references

In this book, we have studied only a very particular class of differential inclusions and some of their applications in Mechanics (dynamical and quasistatical problems in finite-dimensional settings). The purpose of this section is to expand somewhat the reference list, in order to give a fairer view of this research field and to suggest further readings of more or less strongly related subjects. It must be said that we do not try to give a complete list (there is no such thing) and that only a partial search of our files has been made, occasionally indulging in quoting references. By way of excuse, it is hoped that authors not mentioned here are at least cited in some of the works below.

There are several books treating differential inclusions. They deal mainly, if not exclusively, with differential inclusions of the general form

$$(1) \qquad \frac{dx}{dt}(t) \, \varepsilon \, F(t, x(t)), \text{ Lebesgue a. e. },$$

which have absolutely continuous solutions. Of course, they also contain a lot of useful information for the study of the problems presented here. Some of the main references in this context are [Aub-Cel] , [Aub-Fra] , [Cas-Val] , [Dei 1] and [Fil].

Sufficient information on Convex, Functional and Set-Valued Analysis and introductions to evolution problems may be found in the above books and also in [Ber], [Bre 2], [Cla], [Eke-Tem] and [Roc 2].

Sweeping processes by nonconvex sets have also been the subject of study, namely in [Val 4-6] and [Cas & al]. If the moving set is the complement of a convex set, then we may say that the point is pushed by the convex set.

For nonsmooth and not necessarily convex analysis, we may suggest for instance [Cla] and [Pan 2].

We may also consider perturbations of the sweeping process, with equation

$$(2) \qquad -\frac{dx}{dt}(t) \, \varepsilon \, N_{C(t)}(x(t)) + F(t, x(t)) ,$$

where F is a (bounded) closed (convex or nonconvex) valued multifunction. Researchs on these or similar problems include [Bah], [Cel-Mar], [Cel-Sta], [Col & al], [Dau], [Gam 1,2], [Kra-Pap], [Lar] and [Mon 4]. Perturbations of evolution equations have been studied at least since [Bre 1] where for instance the following problem is considered:

$$(3) \qquad -\frac{dx}{dt}(t) \in Ax(t) + f(t, x(t)) ,$$

where A is a maximal monotone operator and f is a Lipschitz-continuous single-valued perturbation. This was extended [Att-Dam 1] to multivalued perturbations, with closed convex values and upper semicontinuous in the state variable x. Further extensions, as in the above works, allow A to be time-dependent (as in (2)) and the multivalued perturbations to be lower semicontinuous.

To go back a little further (but not all the way back), let us recall that much attention has been devoted to evolution equations in Hilbert or Banach spaces, in works such as [Ben], [Ben-Bre], [Bre 1], [Cra], [Kat] and [Kom]. The case of a convex set, or more generally of a domain depending on time, is studied for instance in [Att-Dam 2], [Bir], [Ken-Ota] and [Pera]. Furthermore, books such as [Bar], [Pav] and [Vra] are also available.

Recently, a combination of these problems and of those in Chapter 2 has been studied in [Vla], which considers the following inclusion:

$$(4) \qquad -\frac{dx}{dt}(t) \in A^t x(t) ,$$

where A^t is a set-valued maximal monotone operator in a Hilbert space H. It is assumed that the domain $D(A^t) \subset H$ may depend on the time t and that its interior is nonempty:

$$(5) \qquad \text{int } D(A^t) \neq \emptyset ,$$

similarly to Chapter 2. A distance between maximal monotone operators is introduced which allows the definition of a continuous dependence $t \rightarrow A^t$. Existence of absolutely continuous, cbv or bv solutions to (4) is established under appropriate hypotheses.

The second-order sweeping process of §5.1 is revisited in [Cas & al]. First, existence is shown in the case of a Lipschitz-continuous anti-monotone convex-valued multifunction, without assumption on the interior; in that case, the solution q has an absolutely continuous derivative u and the equation may be written as follows (compare with (1.22))

$$(6) \qquad -\frac{du}{dt}(t) \in N_{F(q(t))}(u(t)) ,$$

Lebesgue a. e.. Then, a new existence proof for the problem of §5.1 is given, which relies on the approximation from the exterior by [Gav] (cf. §5.2).

Second-order differential inclusions appear also in the context of control or stabilization problems. For instance, [Tat] deals with the problem of stabilizing the controlled equation

$$(7) \qquad x'' + \partial \phi(x) \ni u \ , \ \| u(t) \| \leq 1 \ ,$$

by a feedback law $u = \psi(x')$. This has an application to the motion above some rigid obstacle of an elastic string having a discrete distribution of masses.

An interesting parallel development is the study of elliptic equations involving measures [Bre 3] . For instance, we may consider [Att & al] :

$$(8) \qquad -\Delta u + \Gamma(x, u(x)) \ni \mu, \ \text{on} \ \Omega$$

and $u = 0$ on the boundary, where Γ is a (maximal monotone) multifunction and μ is a measure. The problem is solved by approximating the measure μ by L^2 functions and by passing to the limit in the corresponding minimization problems thanks to epiconvergence methods (see [Att]).

The numerical solution of evolution equations such as

$$(9) \qquad -\frac{dx}{dt}(t) \, \varepsilon \ \partial f(t, x(t)) \ ,$$

where the effective domain of $f(t, .)$ may change with t is studied in [Ala].

Approximation of an upper semicontinuous multifunction with convex values Γ by continuous functions f_n in the sense of $e(gr f_n, gr \Gamma) \to 0$ (or $h(gr f_n, gr \Gamma) \to 0$, in special situations) was obtained by Cellina [Cel 1-2], while [Cel 3] deals also with nonconvex cases. A concept of stability for approximations of set-valued maps was introduced in [Aub-Wet], where relations between pointwise and graph convergence of sequences of multifunctions are also provided. We may find more references (such as [DeB-Myj] and [Ole]) in [Aub-Fra].

Differential inclusions depending on a parameter are the subject of [Bon-Mar] and [Nas-Ric], among others.

Stochastic differential equations with reflecting boundary conditions (such as in [Tan]), also called Skorohod problems, are treated in [Fra] and references therein.

For the relation between differential inclusions and optimal control, we may suggest [Bla-Fil] and [Kis].

On friction related problems, the survey [Tel] has an impressive list of almost 300 references.

Periodic solutions to one-dimensional dry friction problems are studied via differential inclusions in [Dei 2].

The numerical solution of friction problems (either for systems with a finite number of degrees of freedom or in the setting of continuum mechanics) has been studied by many authors. Let us mention just a few works more: [Ala-Cur], [Kla] and [Ode-Mar], and refer to [Cur 2] for a better perspective of current research.

An interesting problem is that of a material point moving on a sliding plane (that has a known motion) and subjected to percusssion forces, expressed by measures, and to Coulomb friction. This leads to the consideration of equations with impulses and is studied in [Lag].

The related problem of impulsive control deals for instance with differential equations of the following type:

$$(10) \qquad \dot{x}(t) = f(x(t)) + \sum_{i=1}^{m} g_i(x(t))\, \dot{u}_i(t),$$

where the control function u is just a vector function of bounded variation. This has been studied by several authors among which [Bres 2], [Bres-Ram], [Lak & al] and [Vin-Per].

An interesting feature of problem (10) is that the interpretation of the products $g_i(x(t))\, \dot{u}_i(t)$ which is supplied directly by measure theory is not satisfactory for a well-posed treatment of the control evolution problem, as pointed out in [Haj]. This leads to consider instantaneous evolutions and graph solutions ([Bres 2], [Bres-Ram]).

Scattered in the text, we have mentioned some as yet unsolved problems, which hopefully will interest the reader and receive an answer soon. A lot more may surely be found in the references above.

167

Bibliography

The following abbreviations are used:
CRAS: Comptes Rendus de l'Académie des Sciences de Paris,
USTL: Université des Sciences et Techniques du Languedoc,
SAC: (Travaux du) Séminaire d'Analyse Convexe, USTL,
LMGMC: Laboratoire de Mécanique Générale des Milieux Continus, USTL.
References are ordered alphabetically by the abbreviations of authors' names.

[Ala] P. ALART, *Contribution à la résolution numérique des inclusions différentielles*, Thèse, USTL, Montpellier, 1985.

[Ala-Cur] P. ALART & A. CURNIER, Contact discret avec frottement: unicité de la solution, convergence de l'algorithme, École Polytechnique Fédérale de Lausanne, 1987.

[Ame-Pro] L. AMERIO & G. PROUSE, Study of the motion of a string vibrating against an obstacle, Rend. Mat., 8 (1975) 563-585.

[Att] H. ATTOUCH, *Variational convergence for functions and operators*, Pitman, 1984.

[Att-Dam 1] H. ATTOUCH & A. DAMLAMIAN, On multivalued evolution equations in Hilbert spaces, Israel J. Math., 12 (1972) 373-390.

[Att-Dam 2] H. ATTOUCH & A. DAMLAMIAN, Problèmes d'évolution dans les Hilberts et applications, J. Math. pures et appl., 54 (1975) 53-74.

[Att & al] H. ATTOUCH, G. BOUCHITTÉ & M. MABROUK, Formulations variationnelles pour des équations elliptiques semi-linéaires avec second membre mesure, CRAS, Série I, 306 (1988) 161-164.

[Aub-Cel] J. P. AUBIN & A. CELLINA, *Differential Inclusions*, Springer Verlag, Berlin, 1984.

[Aub-Fra] J. P. AUBIN & H. FRANKOWSKA, *Set-Valued Analysis*, Birkhäuser, Boston, 1990.

[Aub-Wet] J. P. AUBIN & R. J. B. WETS, Stable approximations of set-valued maps, Ann. Inst. Henri Poincaré, 5 (1988) 519-535.

[Bah] S. BAHI, Perturbations semi-continues inférieurement d'un problème d'évolution, SAC, Montpellier, 13 (1983) exposé n° 5.

[Ban] S. BANACH, Sur les opérations dans les ensembles abstraits et leur application aux équations intégrales, Fund. Math., 3 (1922) 133-181.

[Bar] V. BARBU, *Nonlinear semigroups and differential equations in Banach spaces*, Noordhoff Int. Publ., Leyden, 1976.

[Bar-Pre] V. BARBU & T. PRECUPANU, *Convexity and optimization in Banach spaces*, 2nd edition, D. Reidel Pub. Co., Dordrecht, 1986.

[Ben] P. BÉNILAN, *Équations d'évolution dans un espace de Banach quelconque et applications*, Thèse d'État, Université de Paris XI, Orsay, 1972.

[Ben-Bre] P. BÉNILAN & H. BRÉZIS, Solutions faibles d'équations d'évolution dans les espaces de Hilbert, Ann. Inst. Fourier, 22 (1972) 311-329.

[Ber] C. BERGE, *Espaces topologiques. Fonctions multivoques*, Dunod, Paris, 1959 (2nd edition, 1966).

[Bir] M. BIROLI, Sur les inéquations d'évolution avec convexe dépendant du temps, Ricerche Mat., 21 (1972) 17-47.

[Bla-Fil] V. I. BLAGODAT·SKIKH & A. F. FILIPPOV, Differential inclusions and optimal control, Proceedings of the Steklov Institute of Mathematics, 1986, 199-259.

[Bon-Mar] G. BONANNO & S. A. MARANO, Random differential inclusions depending on a parameter, Journal of Mathematical Analysis and Applications, 161 (1991) 35-49.

[Bre 1] H. BRÉZIS, *Opérateurs maximaux monotones et semi-groupes de contractions dans les espaces de Hilbert*, Lecture Notes in Math., North-Holland, Amsterdam, 1973.

[Bre 2] H. BRÉZIS, *Analyse Fonctionelle et Applications*, Masson, Paris, 1983.

[Bre 3] H. BRÉZIS, *Non linear elliptic equations involving measures*, Pitman, 1983.

[Bres 1] Alberto BRESSAN, On differential relations with lower continuous right-hand side. An existence theorem, Journ. Diff. Equa., 37 (1980) 89-97.

[Bres 2] Alberto BRESSAN, On differential systems with impulsive controls, Rend. Sem. Mat. Univ. Padova, 78 (1987) 227-236.

[Bres-Ram] Alberto BRESSAN & F. RAMPAZZO, On differential systems with vector-valued impulsive control, Boll. Un. Mat. Ital. B, 3 (1988) 641-656.

[Bress 1] Aldo BRESSAN, Incompatibilità dei teoremi di esistenza e di unicità del moto per un tipo molto comune e regolare di sistemi meccanici, Ann. Scuola Norm. Sup. Pisa, Ser. III, 14 (1960) 333-348.

[Bress 2] Aldo BRESSAN, On the equilibrium in the presence of friction I, Rendiconti del Circolo Matematico di Palermo, Ser. II, 29 (1980) 435-449.

[Bress 3] Aldo BRESSAN, Questioni di regolarità e di unicità del moto in presenza di vincoli olonomi unilaterali, Rend. Sem. Mat. Univ. Padova, 29 (1959) 271-315.

[Bress 4] Aldo BRESSAN, On the equilibrium in the presence of friction II, Rendiconti del Circolo Matematico di Palermo, Ser. II, 30 (1981) 148-156.

[But-Per 1] G. BUTTAZZO & D. PERCIVALE, Sull'approssimazione del problema del rimbalzo unidimensionale, Scuola Norm. Sup. Pisa, E. T. S. Pisa (1980).

[But-Per 2] G. BUTTAZZO & D. PERCIVALE, The bounce problem on n-dimensional Riemannian manifolds, Scuola Norm. Sup. Pisa, E. T. S. Pisa (1981).

[Cas 1] C. CASTAING, Sur une nouvelle classe d'équation d'évolution dans les espaces de Hilbert, SAC, Montpellier, 13 (1983) exposé n° 10.

[Cas 2] C. CASTAING, Version aléatoire du problème de rafle par un convexe variable, CRAS, Série A-B, 277 (1973) A1057-A1059.

[Cas 3] C. CASTAING, Version aléatoire du problème de rafle par un convexe variable, SAC, Montpellier, 4 (1974) exposé n° 1.

[Cas 4] C. CASTAING, Rafle par un convexe aléatoire à variation continue à droite, SAC, Montpellier, 5 (1975) exposé n° 15.

[Cas 5] C. CASTAING, Rafle par un convexe aléatoire à variation continue à droite, CRAS, Sér. A, 282 (1976) 515-518.

[Cas 6] C. CASTAING, Quelques problémes de mesurabilité liés à la théorie de la commande, CRAS, Sér. A, 262 (1966) 409-411.

[Cas 7] C. CASTAING, Quelques problèmes d'évolution du second ordre, SAC, Montpellier, 18 (1988) exposé n° 5.

[Cas 8] C. CASTAING, Quelques résultats de convergence dans les inclusions différentielles, SAC, Montpellier, 17 (1987) exposé n° 12.

[Cas-Val] C. CASTAING & M. VALADIER, *Convex Analysis and Measurable Multifunctions*, Lecture Notes in Math. 580, Springer Verlag, Berlin, 1977.

[Cas & al] C. CASTAING, T. X. DUC HA & M. VALADIER, Evolution equations governed by the sweeping process, submitted.

[Cel 1] A. CELLINA, Multivalued differential equations and ordinary differential equations, SIAM J. Appl. Math., 18 (1970) 533-538.

[Cel 2] A. CELLINA, Multivalued functions and multivalued flows, Univ. of Maryland Tech. Note BN-615, 1969.

[Cel 3] A. CELLINA, Approximation of set valued functions and fixed point theorems, Annali di Mat. Pura Appl., 82 (1969) 17-24.

[Cel-Mar] A. CELLINA & M. V. MARCHI, Non-convex perturbations of maximal monotone differential inclusions, Israel J. Math., 46 (1983) 1-11.

[Cel-Sta] A. CELLINA & V. STAICU, On evolution equations having monotonicities of opposite sign, SISSA, Trieste, Preprint 142M, 1989.

[Chr] M. CHRAIBI KAADOUD, *Étude théorique et numérique de problèmes d'évolution en présence de liaisons unilatérales et de frottement*, Thèse 3ème cycle, USTL, Montpellier, 1987.

[Cit] C. CITRINI, Controesempi all'unicità del moto di una corda in presenza di una parete, Atti Accad. Naz. Lincei, Rend. Cl. Sci. Fis. Mat. Natur. (8) 67 (1979) 179-185.

[Cla] F. H. CLARKE, *Optimization and Nonsmooth Analysis*, Wiley-Interscience, 1983.

[Col & al] G. COLOMBO, A. FONDA & A. ORNELAS, Lower semicontinuous perturbations of maximal monotone differential inclusions, Israel J. Math., 61 (1988) 211-218.

[Cra] M. G. CRANDALL, *Nonlinear Evolution Equations*, Academic Press, New York, 1979.

[Cur 1] A. CURNIER, A theory of friction, Int. J. Solids Struct., 20 (1984) 637-647.

[Cur 2] A. CURNIER (Ed.), *Proceedings Contact Mechanics International Symposium*, Presses Polytechniques et Universitaires Romandes, Lausanne, 1992.

[Dau] J.-P. DAURÈS, Un problème d'existence de commandes optimales avec liaison sur l'état, CRAS, 279 (1974) 511-514.

[DeB-Myj] F. DE BLASI & J. MYJAK, On continuous approximations for multifunctions, Pacific J. Math., 123 (1986) 9-31.

[Dei 1] K. DEIMLING, *Multivalued differential equations*, W. de Gruyter, Berlin, New York, 1992.

[Dei 2] K. DEIMLING, Multivalued differential equations and dry friction problems, in *Delay and Differential Equations*, (eds. A. M. Fink, R. K. Miller and W. Kliemann), World Scientific Publ., Singapore, 1992.

[Del 1] E. DELASSUS, Mémoire sur la théorie des liaisons finies unilatérales, Ann. Sci. École Norm. Sup., 34 (1917) 95-179.

[Del 2] E. DELASSUS, Considérations sur le frottement de glissement, Nouv. Ann. de Mathématiques, 4ème Série, 20 (1920) 485-496.

[Do] C. DO, On the dynamic deformation of a bar against an obstacle, in *Variational Methods in the Mechanics of Solids* (S. Nemat-Nasser Ed.) Pergamon Press, 1980.

[Dun-Sch] N. DUNFORD & J. T. SCHWARTZ, *Linear Operators - Part I*, Interscience Publishers, New York, 1967 (4th edition).

[Eke-Tem] I. EKELAND & R. TEMAM, *Convex Analysis and Variational Problems*, North-Holland, Amsterdam, 1976.

[Fil] A. F. FILLIPOV, *Differential Equations with Discontinuous Right-hand Sides*, Kluwer Academic Publishers, Dordrecht, 1988.

[Fra] H. FRANKOWSKA, A viability approach to the Skorohod problem, Stochastics, 14 (1985) 227-244.

[Fry] A. FRYSZKOWSKI, Carathéodory-type selectors of set-valued maps of two variables, Bull. Acad. Pol. Sci., 25 (1977) 41-46.

[Gam 1] A. GAMAL, Perturbations semi-continues supérieurement de certaines équations d'évolution, SAC, Montpellier, 11 (1981) exp. n°14.

[Gam 2] A. GAMAL, Perturbation non convexe d'un problème d'évolution dans un espace hilbertien, SAC, Montpellier, 11 (1981) exposé n° 16.

[Gav] A. GAVIOLI, Approximation from the exterior of a multifunction and its applications in the "sweeping process", J. Diff. Equa., 92 (1991) 373-383.

[Haj] O. HAJEK, Book review, Bull. Amer. Math. Soc., 12 (1985) 272-279.

[Jea] M. JEAN, Un algorithme numérique simple pour un système d'oscillateurs avec frottement de Coulomb sur un plan, LMGMC, Montpellier, Note Technique 85-2, 1985.

[Jea-Mor] M. JEAN & J. J. MOREAU, Dynamics in the presence of unilateral contacts and dry friction: a numerical approach, LMGMC, Montpellier, Preprint 85-5, 1985.

[Jea-Pra] M. JEAN & E. PRATT, A system of rigid bodies with dry friction, Int. J. Engineering Sci., 23 (1985) 497-513.

[Jef] R. L. JEFFERY, Non-absolutely convergent integrals with respect to functions of bounded variation, Trans. A.M.S., 34 (1932) 645-675.

[Kat] T. KATO, Nonlinear semigroups and evolution equations, J. Math. Soc. Japan, 19 (1967) 508-520.

[Ken-Ota] N. KENMOCHI & M. ÔTANI, Instability of periodic solutions of some evolution equations governed by time-dependent subdifferential operators, Proc. Japan Acad. Ser. A, 61 (1985) 4-7.

[Kis] M. KISIELEWICZ, *Differential Inclusions and Optimal Control*, P. W. N. , Warsaw and Kluwer Acad. Publ., Dordrecht/Boston/London, 1991.

[Kla] A. KLARBRING, A mathematical programming approach to three-dimensional contact problems with friction, Comp. Meth. Appl. Mech. Eng., 58 (1986) 175-200.

[Kom] Y. KOMURA, Non linear semigroups in Hilbert spaces, J. Math. Soc. Japan, 19 (1967) 493-507.

[Kra-Pap] D. KRAVVARITIS & N. S. PAPAGEORGIOU, Multivalued perturbations of subdifferential type evolution equations in Hilbert spaces, J. Diff. Equa., 76 (1988) 238-255.

[Lag] M. LAGHDIR, *Solution des équations régissant le mouvement de particules en contact avec frottement sec et recevant des impulsions*, Thèse, USTL, Montpellier, 1987.

[Lak & al] V. LAKSHMIKANTHAM, D. D. BAINOV & P. S. SIMEONOV, *Theory of impulsive differential equations*, World Scientific Publ., Singapore, 1989.

[Lar] N. LARHRISSI, Perturbation à valeurs faiblement compactes non nécessairement convexes d'un problème d'évolution, SAC, Montpellier, 14 (1984) exposé n° 14.

[Löt 1] P. LÖTSTEDT, Mechanical systems of rigid bodies subject to unilateral constraints, SIAM J. Appl. Math., 42 (1982) 281-296.

[Löt 2] P. LÖTSTEDT, Coulomb friction in two-dimensional rigid body systems, Z. Angew. Math. u. Mech., 61 (1981) 605-615.

[Löt 3] P. LÖTSTEDT, Numerical simulation of time-dependent contact and friction problems in rigid body mechanics, TRITA-NA-8214, Dept. Numerical Analysis, Royal Institute of Technology, Stockholm.

[Mar & al] J. A. C. MARTINS, M. D. P. MONTEIRO MARQUES & F. GASTALDI, On an example of non-existence of solution to a quasistatic frictional contact problem, European J. Mechanics A/Solids, to appear.

[Mau] S. MAURY, Un problème de frottement équivalent à un problème de poursuite: étude assymptotique, SAC, Montpellier, 4 (1974) exposé n° 10.

[Mic] E. MICHAEL, Continuous selections I, Ann. Math., 63 (1956) 361-381.

[Mon 1] M. D. P. MONTEIRO MARQUES, *Quelques questions d'Analyse et Géométrie posées par la Mécanique des Milieux Continus*, Thèse de 3ème cycle, USTL, Montpellier, 1983.

[Mon 2] M. D. P. MONTEIRO MARQUES, Dualité et sous-differentiabilité de fonctionnelles construites au moyen d'un opérateur divergence, SAC, Montpellier, 13 (1983) exposé n° 1.

[Mon 3] M. D. P. MONTEIRO MARQUES, Sur la frontière d'un convexe mobile, Atti della Acc. Naz. dei Lincei, Rend. Cl. Sci. Fis. Mat. Natur. 77 (1984) 71-75.

[Mon 4] M. D. P. MONTEIRO MARQUES, Perturbations convexes semicontinues supérieurement de problèmes d'évolution dans les espaces de Hilbert, SAC, Montpellier, 14 (1984) exposé n° 2.

[Mon 5] M. D. P. MONTEIRO MARQUES, Regularization and graph approximation of a discontinuous evolution problem, J. Diff. Equa., 67 (1987) 145-164.

[Mon 6] M. D. P. MONTEIRO MARQUES, Rafle par un convexe semicontinu inférieurement d'intérieur non vide en dimension finie, CRAS, Série I, 299 (1984) 307-310; the complete proof is found in:

[Mon 6a] M. D. P. MONTEIRO MARQUES, Rafle par un convexe semicontinu inférieurement d'intérieur non vide en dimension finie, SAC, Montpellier, 14 (1984) exposé n° 6.

[Mon 7] M. D. P. MONTEIRO MARQUES, Rafle par un convexe continu d'intérieur non vide en dimension infinie, SAC, Montpellier, 16 (1986) exposé n° 4.

[Mon 8] M. D. P. MONTEIRO MARQUES, Chocs inélastiques standards: un résultat d'existence, SAC, Montpellier, 15 (1985) exposé n° 4 and LMGMC, preprint n° 85-3, 1985.

[Mon 9] M. D. P. MONTEIRO MARQUES, Approximation methods for a quasistatic twodimensional evolution problem with frictional unilateral constraint, Advances in Mathematical Sciences and Applications, to appear.

[Mor 1] J. J. MOREAU, Evolution problem associated with a moving convex set in a Hilbert space, J. Diff. Equa., 26 (1977) 347-374.

[Mor 2] J. J. MOREAU, Rafle par un convexe variable: première partie, SAC, Montpellier, 1 (1971) exposé n° 15.

[Mor 3] J. J. MOREAU, On unilateral constraints, friction and plasticity, in *New Variational Techniques in Mathematical Physics* (G. Capriz & G. Stampacchia, Ed.) pp. 173-322, C. I. M. E. II Ciclo 1973, Ediz. Cremonese, Roma, 1974.

[Mor 4] J. J. MOREAU, Approximation en graphe d'une évolution discontinue, R. A. I. R. O. Analyse Numérique/Numerical Analysis, 12 (1978) 75-84.

[Mor 5] J. J. MOREAU, Sur les mesures différentielles de fonctions vectorielles et certains problèmes d'évolution, CRAS, Sér. A-B, 282 (1976) A837-A840.

[Mor 6] J. J. MOREAU, Sur les mesures différentielles des fonctions vectorielles à variation localement bornée, SAC, Montpellier, 5 (1975) exposé n° 17.

[Mor 7] J. J. MOREAU, Intersection of moving convex sets in a normed space, Math. Scand., 36 (1975) 159-173.

[Mor 8] J. J. MOREAU, Un cas de convergence des itérées d'une contraction d'un espace hilbertien, CRAS, Sér. A, 286 (1978) 143-144.

[Mor 9] J. J. MOREAU, Rafle par un convexe variable (deuxième partie), SAC, Montpellier, 2 (1972) exposé n° 3.

[Mor 10] J. J. MOREAU, Bounded variation in time, in *Topics in Non-smooth Mechanics* (J. J. Moreau, P. D. Panagiotopoulos, G. Strang, Eds.), Birkhäuser Verlag, Basel-Boston-Berlin, 1988.

[Mor 11] J. J. MOREAU, Standard inelastic shocks and the dynamics of unilateral constraints, in *Unilateral Problems in Structural Analysis*, (G. del Piero & F. Maceri, Eds.), C.I.S.M. Courses and Lectures n° 288, Springer Verlag, Wien, New York, 1985, 173-221; and also LMGMC, Montpellier, Preprint n° 84-2, 1984.

[Mor 12] J. J. MOREAU, Dynamique de systèmes à liaisons unilatérales avec frottement sec éventuel: essais numériques, LMGMC, Montpellier, Note Technique n° 85-1, 1985.

[Mor 13] J. J. MOREAU, Une formulation du contact à frottement sec; application au calcul numérique, CRAS, Sér. II, 302 (1986) 799-801.

[Mor 14] J. J. MOREAU, Liaisons unilatérales sans frottement et chocs inélastiques, CRAS, Sér. II, 296 (1983) 1473-1476.

[Mor 15] J. J. MOREAU, Quadratic programming in mechanics: dynamics of one-sided constraints, SIAM J. Control, 4 (1966) 153-158.

[Mor 16] J. J. MOREAU, One-sided constraints in hydrodynamics, in *Nonlinear Programming* (J. Abadie, Ed.) pp. 259-279, North-Holland, Amsterdam, 1967.

[Mor 17] J. J. MOREAU, Application of convex analysis to the treatment of elastoplastic systems, in *Applications of Methods of Functional Analysis to Problems of Mechanics (Joint Symposium IUTAM/IMU, Marseille 1975)* (P. Germain & B. Nayroles Eds.) Lecture Notes in Math. 503, Springer Verlag, 1976, pp. 56-89.

[Mor-Val 1] J. J. MOREAU & M. VALADIER, Dérivation d'une mesure vectorielle sur un intervalle, SAC, Montpellier, 14 (1984) exposé n° 1.

[Mor-Val 2] J. J. MOREAU & M. VALADIER, Quelques résultats sur les fonctions vectorielles à variation bornée d'une variable réelle, SAC, Montpellier, 14 (1984) exposé n° 16.

[Mor-Val 3] J. J. MOREAU & M. VALADIER, A chain rule involving vector functions of bounded variation, J. Funct. Anal., 74 (1987) 333-345; and LMGMC, Montpellier, Preprint n° 86-1, 1986.

[Nas-Ric] O. NASELLI RICCERI & B. RICCERI, Differential inclusions depending on a parameter, Bull. Polish Acad. Sci. Math., 37 (1989) 7-12.

[Ode-Mar] J. T. ODEN & J. A. C. MARTINS, Models and computational methods for dynamic friction phenomena, Comp. Meth. Appl. Mech. Eng., 52 (1985) 527-634.

[Ole] C. OLECH, Approximation of set-valued functions by continuous functions, Colloq. Math., 19 (1968) 285-293.

[Pan 1] P. D. PANAGIOTOPOULOS, Non-convex superpotentials in the sense of F. H. Clarke and applications, Mech. Res. Comm., 8 (1981) 335-340.

[Pan 2] P. D. PANAGIOTOPOULOS, *Inequality Problems in Mechanics and Applications*, Birkhäuser, Boston, Basel, Stuttgart, 1985.

[Pav] N. H. PAVEL, *Nonlinear evolution operators and semigroups*, Lecture Notes in Math. n° 1260, Springer, New York, 1987.

[Per] D. PERCIVALE, Uniqueness in the elastic bounce problem, J. Diff. Equa., 56 (1985) 206-215; 2nd part, J. Diff. Equa., 90 (1991) 304-315.

[Pera] J. C. PERALBA, Un problème d'évolution relatif à un opérateur sous-différentiel dépendant du temps, CRAS, 275 (1972) 93-96.

[Rio 1] H. RIOS, *Étude de certains problèmes paraboliques: existence et approximation des solutions*, Thèse de Doctorat de Spécialité, USTL, Montpellier, 1977.

[Rio 2] H. RIOS, Une étude d'existence sur certains problèmes paraboliques, Annales Faculté des Sciences Toulouse, 1 (1979) 235-255.

[Roc 1] R. T. ROCKAFELLAR, Integral functionals, normal integrands and measurable selections, in *Nonlinear Operators and the Calculus of Variations* (Gossez, Lami Dozo, Mawhin and Waelbroeck, Eds.), Lecture Notes 543, Springer Verlag, Berlin, 1976, pp. 157-207.

[Roc 2] R. T. ROCKAFELLAR, *Convex Analysis*, Princeton Univ. Press, Princeton, 1970.

[Rol] S. ROLEWICZ, On the minimum time control problem and continuous families of convex sets, Studia Math., 56 (1976) 39-45.

[Sch 1] M. SCHATZMAN, Le système différentiel $d^2u/dt^2 + \partial\phi(u) \ni f$ avec conditions initiales, CRAS, Sér. A, 284 (1977) 603-606.

[Sch 2] M. SCHATZMAN, A class of non linear differential equations of second order in time, J. Nonlinear Analysis, Theory, Methods and Applications, 2 (1978) 355-373.

[Sch 3] M. SCHATZMAN, *Problèmes unilatéraux d'évolution du 2ème ordre en temps*, Thèse de Doctorat d'État, Univ. Pierre et Marie Curie, Paris, 1979.

[Tan] H. TANAKA, Stochastic differential equations with reflecting boundary conditions in convex regions, Hiroshima Math. Journ., 9 (1979) 163-177.

[Tat] D. TATARU, Stabilizing second order differential equations, Differential and Integral Equations, 2 (1989) 132-143.

[Tau 1] K. TAUBERT, Differenzverfahren für Schwingungen mit trodkener und zäher Reibung und für Regelungssysteme, Numer. Math., 26 (1976) 379-399.

[Tau 2] K. TAUBERT, Converging multistep methods for initial value problems involving multivalued maps, Computing, 27 (1981) 123-136.

[Tay] H. E. TAYLOR, Grazing rays and reflections of singularities, Comm. Pure Appl. Math., 28 (1975) 1-38.

[Tel] J. J. TELEGA, Topics on unilateral contact problems of elasticity and inelasticity, in *Nonsmooth Mechanics and Applications* (J. J. Moreau, P. D. Panagiotopoulos, Eds.), CISM Courses and Lectures n° 302, Springer Verlag, Wien/New York, 1988, p. 340-461.

[Val 1] M. VALADIER, Lipschitz approximation of the sweeping (or Moreau) process, J. Diff. Equa., 88(1990)248-264.

[Val 2] M. VALADIER, Quelques résultats de base concernant le processus de rafle, SAC, Montpellier, 18 (1988) exposé n° 3.

[Val 3] M. VALADIER, Approximation lipschitzienne par l'intérieur d'une multifonction s.c.i., SAC, Montpellier, 17 (1987) exposé n° 11.

[Val 4] M. VALADIER, Quelques problèmes d'entraînement unilatéral en dimension finie, SAC, Montpellier, 18 (1988) exposé n° 8.

[Val 5] M. VALADIER, Lignes de descente de fonctions lipschitziennes non pathologiques, SAC, Montpellier, 18 (1988) exposé n° 9.

[Val 6] M. VALADIER, Entraînement unilatéral, lignes de descente, fonctions lipschitziennes non-pathologiques, CRAS, Série I, 308 (1989) 241-244.

[Vin-Per] R. B. VINTER & F. M. F. L. PEREIRA, A maximum principle for optimal processes with discontinuous trajectories, SIAM J. Control and Optimization, 26 (1988) 205-229.

[Vla] A. A. VLADIMIROV, Nonstationary dissipative evolution equations in a Hilbert space, Nonlinear Analysis, Theory, Methods and Applications, 17 (1991) 499-518.

[Vra] I. I. VRABIE, *Compactness methods for nonlinear evolutions*, Pitman Monographs and Surveys in Pure and Appl. Math. 32, Longman, 1987.

Index

Index of Notation

Progress in Nonlinear Differential Equations
and Their Applications

Editor
Haim Brezis
Département de Mathématiques
Université P. et M. Curie
4, Place Jussieu
75252 Paris Cedex 05
France
and
Department of Mathematics
Rutgers University
New Brunswick, NJ 08903
U.S.A.

Progress in Nonlinear and Differential Equations and Their Applications is a book series that lies at the interface of pure and applied mathematics. Many differential equations are motivated by problems arising in such diversified fields as Mechanics, Physics, Differential Geometry, Engineering, Control Theory, Biology, and Economics. This series is open to both the theoretical and applied aspects, hopefully stimulating a fruitful interaction between the two sides. It will publish monographs, polished notes arising from lectures and seminars, graduate level texts, and proceedings of focused and refereed conferences.

We encourage preparation of manuscripts in some form of TeX for delivery in camera-ready copy, which leads to rapid publication, or in electronic form for interfacing with laser printers or typesetters.

Proposals should be sent directly to the editor or to: Birkhäuser Boston, 675 Massachusetts Avenue, Cambridge, MA 02139

PNLDE 1   Partial Differential Equations and the Calculus of Variations, Volume I
Essays in Honor of Ennio De Giorgi
F. Colombini, A. Marino, L. Modica, and S. Spagnolo, editors

PNLDE 2   Partial Differential Equations and the Calculus of Variations, Volume II
Essays in Honor of Ennio De Giorgi
F. Colombini, A. Marino, L. Modica, and S. Spagnolo, editors

PNLDE 3   Propagation and Interaction of Singularities in Nonlinear Hyperbolic Problems
Michael Beals

PNLDE 4   Variational Methods
Henri Berestycki, Jean-Michel Coron, and Ivar Ekeland, editors

PNLDE 5   Composite Media and Homogenization Theory
Gianni Dal Maso and Gian Fausto Dell'Antonio, editors

Monographs in Mathematics

Managing Editors:

H. Amann (Universität Zürich)
K. Grove (University of Maryland, College Park)
H. Kraft (Universität Basel)
P.-L. Lions (Université de Paris-Dauphine)

Associate Editors:

H. Araki (Kyoto University)
J. Ball (Heriot-Watt University, Edinburgh)
F. Brezzi (Università di Pavia)
K.C. Chang (Peking University)
N. Hitchin (University of Warwick)
H. Hofer (Universität Bochum)
H. Knörrer (ETH Zürich)
K. Masuda (University of Tokyo)
D. Zagier (Max-Planck-Institut, Bonn)

Recently published in the series Monographs in Mathematics:

Volume 82: **V. I. Arnold/S. M. Gusein-Zade/A. N. Varchenko, Singularities of Differentiable Maps - Volume I.**
1985, 392 pages, hardcover, ISBN 3-7643-3187-9.

Volume 83: **V. I. Arnold/S. M. Gusein-Zade/A. N. Varchenko, Singularities of Differentiable Maps - Volume II.**
1988, 500 pages, hardcover, ISBN 3-7643-3185-2.

Volume 84: **Hans Triebel, Theory of Function Spaces II.**
1992, 380 pages, hardcover, ISBN 3-7643-2639-5.

Volume 85: **K.R. Parthasarathy, An Introduction to Quantum Stochastic Calculus.**
1992, 380 pages, hardcover, ISBN 3-7643-2697-2.

Volume 86: **Masao Nagasawa, Schrödinger Equations and Diffusion Theory.**
1993, 332 pages, hardcover, ISBN 3-7643-2875-4.

Volume 87: **Jan Prüss, Evolutionary Integral Equations and Applications.**
1993, 392 pages, hardcover, ISBN 3-7643-2876-2.

MIX
Papier aus verantwortungsvollen Quellen
Paper from responsible sources
FSC® C105338

If you have any concerns about our products,
you can contact us on
ProductSafety@springernature.com

In case Publisher is established outside the EU,
the EU authorized representative is:
**Springer Nature Customer Service Center GmbH
Europaplatz 3, 69115 Heidelberg, Germany**

Printed by Libri Plureos GmbH
in Hamburg, Germany